The
Laboratory
Animal —
Principles and
Practice

The Laboratory Animal— Principles and Practice

W. LANE-PETTER

A. E. G. PEARSON

1971

 Academic Press · London and New York

ACADEMIC PRESS INC. (LONDON) LTD
Shipton House
24–28 Oval Road
London NW1 7DD

U.S. Edition published by
ACADEMIC PRESS INC.
111 Fifth Avenue
New York, New York 10003

Library of Congress Catalog Card Number: 70-17053
International Standard Book Number: 0-12-435-760-1

Printed in Great Britain by
Butler and Tanner Ltd
Frome and London

Preface

This book is, in part, a second edition, albeit much enlarged, of a monograph written by one of us and published ten years ago (Lane-Petter, 1961 : see Bibliography). A few sections of the original monograph have been brought up to date and used again, but for the most part the present book is entirely original. There has been some division of labour between the authors, but the book is very much a collaborative effort, and the authors take joint responsibility for it as a whole.

The book is an outline of the principles governing the breeding, procurement, accommodation, maintenance, distribution and use of laboratory animals. As such, it is addressed to those who have to provide facilities for laboratory animals : the planner, administrator, departmental head, architect and financial controller. They will find here some guidance on matters that may seem elementary to them, but they will also find much practical detail which, while perhaps not of direct interest to them, is certainly not irrelevant. The practical detail, often included at some length because it is not readily accessible elsewhere, is of more moment to those who have to run laboratory animal facilities; but they should not be entirely ignorant of the principles on which their work is based.

It seemed, therefore, logical that an outline of principles and a practical guide should be combined in the same book. Laboratory animal science is not a science, but a transdisciplinary consideration of one of the ground materials of biomedical research, the laboratory animal. There is a danger that those who work with, say, germfree animals will overlook some important social factors that loom large in open colonies; or that the toxicologist will be oblivious of the practical facts of breeding and make impossible or unreasonable demands on his supplier. There is an even greater danger that the study of laboratory animals, which has come so far in the last twenty years, should be thought to be approaching the end of the journey instead of being still near the beginning :

"For there is good news yet to hear and fine things to be seen" and one of the objects of this book is to suggest the directions in which it may be most profitable to look.

Laboratory animals are not chemical reagents, although there are certain common analogies; nor are they little men, whose reactions predict in miniature how man will react, although they are useful, indispensable, indicators. They are *sui generis*, though still of this world, and so long as this is remembered they will continue to contribute irreplaceably, as they have in the past, to our knowledge of ourselves and our biological environment. To get the best out of animal experiments it is necessary to understand the animal, and to understand the animal it is necessary to respect it. The authors hope that their work will enhance the reader's respect for biological science's best friend.

W. LANE-PETTER   *October 1971*   A. E. G. PEARSON
Houghton, Huntingdon   Welwyn Garden City,
                                                  Hertfordshire

Foreword

In these days the name of a branch of the biological sciences is often preceded by the word "molecular". A fact of fundamental importance which is then liable to be forgotten is that however far the molecular study may have moved away from recognizable dependence upon an intact living organism, primary observations involving living cells were the genesis—the *fons et origo*—of the ultimately molecular study. The study of plants, animals, microorganisms—all manifestations of life—is the basis of biological investigation, however remote from intact plants, animals or microorganisms the study may have progressed.

The importance of laboratory animals is immense for the development of physiology, biochemistry, pharmacology, pathology, and of experimental and diagnostic medicine, as well as for the fundamental comparative investigations of zoology and of ecology in general. Nevertheless the study of the conditions under which laboratory animals have been reared, housed and kept has sometimes not received the attention that it deserves, and expensive biological investigations have sometimes failed to achieve a conclusive goal largely because of the inconsistency of the starting material.

Dr Lane-Petter and Dr Pearson each brings to bear on the subject of this book a wide experience of the needs of biological research in industry and in government-financed research institutes. They discuss the importance of the genetics of the laboratory animal, the possibility of finding substitutes for laboratory animals, the health, the environment and the nutrition of experimental animals, and the administrative duties of those concerned with the production of animals for use in laboratory investigations. The growing realization of the need for the control of the human environment has been anticipated by those who know their job, as the authors do, in the production and care of laboratory animals.

In considering possible substitutes for laboratory animals the authors look forward to a time when developments in chemical and biochemical analysis, and in *in vitro* methods, have reduced to a negligible level the need for the use of laboratory animals for the purposes of diagnosis and of mandatory assay. But this situation can be reached only by the use of parallel investigations on animals and on the other systems. And new subjects of investigation will continue to demand the use of animals, during the early stages at least.

The claim is sometimes made that insulin could have been discovered without the use of animals for the testing of the anti-diabetic activity of extracts of the pancreas. What nonsense! In fact Dr Lydia Dewitt undoubtedly extracted insulin from pancreatic islets in 1906 by a method similar to that used in 1921 by Banting and Best. But her investigations were pursued in an Institute of Histology, and perhaps experimentally diabetic animals were not available for the testing of her extract. At any rate she tried out her precious material on an *in vitro* system which did not respond, and in the light of subsequent knowledge would not be expected to respond, to insulin. Had Dr Dewitt injected her extract into depancreatized dogs insulin might have been discovered fifteen years earlier than in fact it was generally recognized. And who knows how many lives would then have been saved of those who died from diabetes mellitus during these fifteen years?

Dr Lane-Petter and Dr Pearson have together written a work which is of great practical importance and which will be consulted by biologists of many different sorts in the years to come. And with the authoritative advice thus available about the choice and treatment of experimental animals, how many investigators will help to make discoveries of importance in the fight against disease both in animals and man? Time will show.

August 1971

F. G. YOUNG, M.A., D.SC., F.R.S.,
Sir William Dunn Professor of Biochemistry, and
Master of Darwin College, University of Cambridge,
Cambridge, England.

Contents

CHAPTER 1

Utilization

I. Introduction

The use of animals in the furtherance of scientific investigation goes back to some five centuries before the dawn of the Christian era. The use was then probably sporadic and limited to man's domesticated commensals. Little investigation was carried out in the scientific field between that time and the seventeenth century, when a burst of activity occurred which subsequently died down until the nineteenth century.

Up to the end of the last century experiments using living animals were carried out on domestic or easily captured wild species; the choice was again limited and casual. By the end of the nineteenth century the laboratory animal was beginning to emerge as a creature in its own right; the science of biology had already made considerable strides outside the confines of a descriptive science into the open country of experimentation. The concept of the laboratory animal had begun to emerge as an animal deliberately chosen for its inherent suitability for the purpose; especially bred in captivity and in the process more or less profoundly modified, or procured from its environment not merely on grounds of convenience but for its fitness in the context of the investigation.

1

It is the purpose of this book to present to the reader the principles of managing these animals, the pitfalls that may be encountered and the economic aspects of undertaking their supervision and care.

II. Distribution of Species

Domesticated animals were the first to be brought under the concept of a laboratory species, but scientists began soon to pay attention to un-domesticated animals, particularly those which, uninvited, had learned to share his immediate environment, namely rats and mice. Rabbits and guinea-pigs were also found to accept captivity in the laboratory environment.

The apparent adaptability of rats and mice made them an ideal choice for breeding and maintenance in captivity. This characteristic resulted in their successful development as species of choice in many laboratories. They were omnivorous, robust, prolific, small and, after a few generations of breeding in captivity, tame. They were therefore cheap and easy to breed and maintain and, in not being highly specialized, were useful in research relevant to human functions and diseases.

These small rodents took their place in the forefront of laboratory animals, a position they hold to the present day. A survey conducted in British laboratories in 1956 (Report, 1958), to identify the relative proportions of species used, demonstrates this preponderance; the results are presented in diagrammatic form in Fig. 1.1. They had come to exploit the man-made environment and resources for their own ends. Unlike most wild animals, they could tolerate the proximity of man; indeed, they sought it for the sake of all the perquisites that went with it, and it provided them with a rich living. It was a natural step for man to take, to invite into his laboratories, as collaborators in his researches, those who throughout the ages had been scratching at his door. As vermin they were and are detested, and with good reason. When they crossed the laboratory threshold and put on white coats they became respectable, pampered and indispensable. From these two species have been developed special strains with characteristics which simplify certain aspects of biological research and which enable results to be more meaningful; inbred strains with high incidences of various spontaneous malignant conditions, strains with different growth characteristics for different purposes, strains which are hairless.

Although rats and mice have held the dominant position in biological research since the dawn of experimental science, it is the guinea-pig that has become synonymous, in the mind of the lay public, with experimental medical research. Perhaps this is due to a subconscious rejection of rats and mice as vermin and the social acceptability of a species frequently employed as a domestic pet. In practice the experimental use of guinea-pigs has always lagged behind the other two species. Its origins are in South

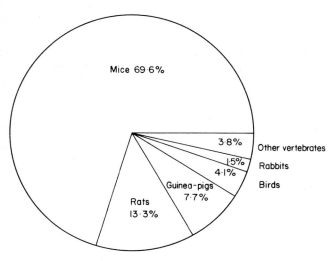

Fig. 1.1. Proportions of vertebrate species used in British laboratories, in 1956.

America where it had already been domesticated; possessing the advantages of being small, robust, reasonably prolific and easily handled, it is a natural species to select. Its failure to achieve the same status as rats and mice is a reflection of its slower and less prolific reproductive capacity and more exacting dietary requirements. It has special advantages in certain research fields in that, like man, it is susceptible to tuberculosis and requires ascorbic acid (vitamin C) in the diet.

The main reasons why rabbits came to be popular in the laboratory are first because of their convenient size when it comes to dissection and, second, because they had become established as a very common domestic animal, used for fur, flesh and fancy, and were therefore easily and cheaply obtainable. They have a particular place in the experimental field due to their ability to allow the growth of adults of certain parasitic helminths, and more recently they have become an essential species in testing for possible teratogenic effects in new therapeutic compounds. However, these last two considerations notwithstanding, the rabbit, though a prolific and adaptable vegetarian, is much more specialized and differs from the rodents in belonging to one of the less successful evolutionary branches, the *Lagomorpha*, and has far less claim to a place in the scientific bestiary than the strikingly successful *Rodentia*.

The use of poultry is mostly confined to experimentation on diseases of its own species, although it has a place in cancer research in the study of virus-induced tumours. Many other vertebrate species are used, but these are restricted to specialized purposes, for which a particular species has peculiar and relevant characteristics, or the need to compare experimental data in different species in order to be satisfied that the results are not narrowly specific. Mention should be made of the use of primates in research related to the human condition; the reasons for this choice are obvious, but primates are difficult to handle, have specialized dietary and environmental requirements, they may carry infectious organisms pathogenic to man, and they are extremely costly. Except in behavioural studies, where they have proved of great value, there appears to be little advantage in their use over other, less hominoid species.

The most comprehensive reference book on this subject is undoubtedly *The UFAW Handbook on the Care and Management of Laboratory Animals*, in which upwards of fifty different vertebrate species are described, most of them in extensive detail. However, this approach tends to obscure the enormous numerical preponderance of mice and rats and, to a lesser extent, guinea-pigs. Nothing succeeds like success and the history of the development of small rodents as predominant laboratory animals is an example of this principle. The more a species is used, the more technical information is collected, from the aspects of both husbandry and physiology, and therefore the greater are the chances that it will be selected for future research. An alternative will only be sought for the sake of some particular characteristic which suits it for a specific experimental purpose. The reasons for selecting the unfamiliar or less defined species will have to be great to outweigh the practical disadvantages of having to undertake preliminary work on adaptation, husbandry, breeding and even physiology. Numbers, for all that, are not the whole story, and a proper perspective can only be obtained by a consideration of the uses to which different species are put. This aspect is extensively reviewed by Russell and Burch (1959).

The position of dogs and cats as laboratory animals should receive special consideration at this point, since their predominance as domestic pets introduces emotional factors which cannot and must not be ignored. The value of these species as test animals in carrying out certain experimental procedures is very great. They possess the advantages of size for carrying out complex surgical and monitoring procedures; their adaptability and easy relationship with man simplifies handling and the interpretation of behaviour; in many physiological situations their responses are very close to those of man. The decision to use these species is not taken lightly, since they are difficult to breed and maintain and the requirement for special accommodation involves large areas and is extremely costly. Although the numbers used are very small when compared with other species, they command a major proportion of the considerations of experimenters and those concerned with laboratory animals, not only for the reasons outlined but also for the emotional and humanitarian aspects. The use of cats and dogs in experimentation requires special permission from the Home Office; this additional protection is offered to no other species with the exception of the *Equidae*.

III. Purposes for which Laboratory Animals are Used

Laboratories using animals may be divided into groups according to the type of work that constitutes their predominant interest. Thus, universities are primarily engaged in fundamental research into the processes of bio- logical activity, industrial laboratories are concerned with routine product testing and monitoring or the discovery of therapeutic agents for the treat- ment of disease in man and animals, and hospital research units are con- cerned with the whole range of medical science. Alternatively, the use of laboratory animals can be classified under purpose headings, regardless of the type of laboratory engaged in such activity. The latter approach is more likely to provide useful information since many laboratories, although having a predominant purpose, engage in a variety of activities.

A. RESEARCH

Under this heading cancer research will be excluded and considered else- where, since the requirements are specialized and extensive. The type and quantity of animal used depends broadly on the classification of research. The concept of pure and applied research is valid in the biological sphere as in the physical sciences. Pure research into fundamental biological processes and the aetiology of disease usually requires the use of small numbers of animals within a wide variety of species selected for characteristics relevant to the research field. This segment of research usually involves complex monitoring by means of sophisticated instrumentation and requires the use of the larger species. Such experiments are demanding in the time taken to to set them up and in their duration, consequently fewer animal numbers are employed. At the other end of the scale, applied research into, for example, the search for therapeutically active compounds requires the use of large numbers of small animals used in screening tests designed to answer a single simple question. The dividing line between these two extremes is not usually so easy to define; much applied research is a combination of both approaches and the two interact during the whole course of research even when broadly described as applied.

B. CANCER RESEARCH

Investigations into the aetiology, prevention and treatment of cancer depend initially on the setting up of a suitable model in the experimental animal. Cancer takes many forms and what is applicable to one type of tumour does not necessarily hold for another. The requirement is therefore for a wide variety of malignant tumours capable of being transplanted from one animal to another. Since malignant tissue may act in the same way as other foreign tissue in producing a general immune response in the host, successful transplantability depends on genetic compatibility between donor and recipient. This requirement led to the development of inbred strains of mice, this species being chosen for its size, rapid breeding and high incidence of a variety of spontaneous neoplasms. Thus, in the sphere of cancer research the mouse holds pride of place; in the 1956 survey (Fig. 1.1) it was shown that 60 per cent of animals used in cancer research were mice, the bulk of the remainder being rats, with much lesser use of chickens, hamsters and an assortment of other species.

C. TOXICOLOGY

The introduction of a new chemical subsistance for consumption by man or domestic animals—be it a food additive for colouring or flavouring, or a drug used in the prevention or treatment of disease—requires prior testing to ensure that no harmful effects will arise, or that the level of side-effects are in balance with the seriousness of the disease being treated. The last two decades have seen a tremendous expansion in the use of man-made compounds in medicine, the food industry and the control of pests. With this expansion has come the realization that desired activity may also have unwanted effects as its inevitable running partner, and consequently this has led to the growth of the science of toxicology. Compounds for use in domestic animals can be tested in these species for these side-effects, but for use in man suitable animals have to be used in an attempt to extrapolate the effects to man. Obviously, the animal of choice must be comparable to man in physiological response and adaptation, but no ideal species exists which completely fills the specification. Man is a genetically variable animal and even within his own species the response to a substance may be very variable. The requirement is therefore for quantity as well as for considerations of physiological approximation. While the use of primates has often been considered, as a choice depending on close evolutionary relationship, the problems associated with their husbandry, handling and housing in large quantities make this group unsuitable for normal purposes. The necessity to test in more than one species is of paramount importance in order to reduce the possibility of a species specific reaction. The rat is usually the animal of choice for preliminary testing in large numbers over a wide range of dose levels. The second species chosen is usually the dog because of its closer approximation to man, ease of handling and the ability on the part of its handlers to detect distress at an early stage. The strain most often employed, for many years, is the beagle. This strain has a very amenable temperament, can be housed in packs, is adaptable to conditions of captivity and is not too large. This latter consideration is important, since feeding costs are high and the availability of the substance to be tested may not be extensive. These characteristics led to an early preference for the beagle and considerable background data have been accumulated, and these in turn provide additional advantages in the selection of this strain. However, this narrow concentration may not be advisable in the light of species or strain specific reactions, and it has been said that the purpose of toxicology is to make the world safe for laboratory rats and beagles—but that is a problem for toxicologists to resolve.

The establishment of a correlation between the administration of the drug thalidomide and the birth of babies with limb deformities led to extensive testing of drugs for teratogenic effects when administered to pregnant animals. Since in the case of thalidomide it was found that administration of the drug to pregnant rabbits produced effects comparable to those found in man, this species is now widely used in toxicology. However, the effect is extremely strain dependent and again the possibility of a unique response raises doubts as to the validity of using this species for universal teratogenic testing. As in the case of general toxicity testing, the rat is tending to become the main species in teratogenic studies.

D. DIAGNOSIS

This heading includes the use of animals for the detection of disease or pregnancy—although this last category is now mainly extinct since the development of hormone assay techniques. The method involves the inoculation of material from the patient into the test animal, and observation of the outcome by post-mortem examination. Sensitivity to the organism concerned on the part of the test animal is therefore mandatory. The diagnosis of tuberculosis and brucellosis in man and cattle accounts for the bulk of the use under this heading, and the animals concerned are guinea-pigs. In the 1956 survey it was found that about half the guinea-pigs used in Great Britain were for this purpose. The test animals must, of course, be free from disease before inoculation and this is not always so easy to achieve; it is therefore inevitable and desirable that laboratory techniques not involving the use of animals will eventually eliminate this category altogether.

The quantities of animals used for diagnostic purposes can be accurately determined by reference to the annual Home Office returns under the Act of 1876 for Great Britain. An analysis of these returns is presented in Table 3.1 (p. 34), where it can be seen that diagnosis comprises only 5 per cent of the total animal usage.

E. BIOASSAY

Many substances administered to man for prophylactic or therapeutic purposes have to be assayed on the living animal, often as a legal requirement, before they may be offered for clinical use. This procedure of bioassay includes the titration of antibodies in immune sera, the estimation of the strength of solutions of biological substances in terms of standardized units (for example, insulin), the detection of pyrogens in solutions intended for injection, the potency of vaccines, and many others. These control procedures are laid down by statutory requirements where such products are to be used in man, and involve the use of large numbers of mice, rats, guinea-pigs and rabbits as well as smaller numbers of other species.

The numbers of animals used for bioassay purposes in Great Britain is considerable when compared with total usage, as seen in the analysis of Home Office returns presented in Table 3.1 (p. 34). The proportion used for this purpose in 1969 was 27 per cent compared with 33 per cent in 1960; this reduction is a reflection of the increased usage for other purposes as it can be seen that the actual numbers used for assay remains reasonably constant.

F. TEACHING

Discounting the animals used for dissection, the numbers used for demonstration purposes are extremely small. The species employed are mainly cats and dogs, for instruction purposes in physiology and pharmacology. While not falling strictly into the category of an experimental animal, the use of animals for teaching purposes in Great Britain is still controlled by the Home Office under the Act of 1876 (p. 255).

IV. Users of Laboratory Animals

It has already been pointed out that a classification of laboratories, by the main purpose to which they put their animals, is liable to be artificial and misleading. However, it is still worth considering some of the main categories of these laboratories.

Four user groups can be identified by consideration of the number of species used and the proportional usage of each.

1. Hospitals and Public Health Laboratories

Small numbers of animals used at a relatively constant rate with species variation low. Mainly guinea-pigs but also mice and rabbits.

2. Pharmaceutical Industry

Constant high-volume usage of rats and mice in advanced research projects or screening procedures; reasonably constant use of smaller numbers of dogs and rats in toxicity testing; irregular use of a variety of species in fundamental research.

3. Cancer Research

Constant high-volume usage of mice with scattered usage of other species.

4. Universities and Research Institutes

Irregular use of a variety of species. These variations in usage pattern have a direct bearing on the prediction of short-term trends. This aspect will be discussed in Chapter 3.

Much of the discussion in this chapter has related to the situation in Great Britain, where a generous distribution of industrial, university and research institute laboratories exists. For the most part these proportions hold good for other countries; however, variations do exist as shown by the ICLA international surveys. These variations are related more to function than to personal preferences; for instance, the high proportion of guinea-pig usage in Turkey (41 per cent of total animals used) was related to an extensive public health campaign against tuberculosis and brucellosis in which the guinea-pig is the most important diagnostic animal. Variations in other countries can usually be attributed to different emphases regarding purpose. However, national preferences over species used for research do exist; Scandinavian countries have a bias towards the use of guinea-pigs, while France, Italy and Austria show a relatively high proportion of rabbits used.

Thus, there is a fairly constant pattern in the choice of animal species throughout the world, which is independent of the volume of work carried out. Variations reflect qualitative differences in emphasis. The suggestion that the choice of the preponderant species is due to tradition, fashion or lack of enterprise does not stand, since science is susceptible to these influences only to a very slight degree. The criteria for species selection depend on the prime requisites for a laboratory animal—that it should be robust, prolific, small and tame. Only where special characteristics have to be sought will unusual species be selected. Pioneer research may often have to go outside the normal run of laboratory animal species, but there is a strong tendency to revert to the more familiar.

CHAPTER 2

Sources

I. Introduction

Among the variety of different species of animals used in the laboratory, some are specially bred for the purpose, some are bred for other purposes and incidentally supply laboratory needs, and some are collected from the wild or the quasi-wild state.

Collected animals include most primates, together with a number of species that are used in only small numbers in the laboratory, and which do not breed in captivity or only with difficulty. Apart from primates, such as macaque monkeys and certain common African and South American species, none of the collected species can be used in large numbers; if they were, they would raise a problem of conservation, because of the probable depletion of natural populations. Indeed, the depletion of wild populations of many primates, like the rhesus monkey which has been so common in India as to be an agricultural pest, is now a matter of serious concern to conservationists, many of whom hold the view that collection will have to be restricted or prohibited altogether in the near future. The collection of certain primate species, whose existence in the wild is threatened, is already prohibited (IUCN, 1966), and if collection of other species is not brought under effective control, they too may be added to the list of threatened species, and their collection prohibited.

A special case of collection is the transfer to laboratories of stray dogs and cats which have no home, no owner and are destined to be destroyed. Every large town or city is liable to have a stray dog and cat problem. Humane and animal welfare societies are in principle opposed to this transfer, and sometimes their opposition is loud and furious. In Great Britain, it is illegal for the police, who have the responsibility of dealing with stray dogs—but not cats—to hand strays over for laboratory use, and they equally decline to hand over stray cats. In the USA and in many other countries it is common practice for stray dogs and cats to be impounded, and if not claimed to be offered to laboratories.

Collected animals may also include those that are bred for purposes other than laboratory use, but of which a small proportion are available for laboratory use when needed. Such are farm animals—cattle, sheep, pigs, goats and horses—and also poultry and rabbits, which are raised mostly for food, and in much larger numbers than would ever be required in the laboratory.

However, the greatest number of laboratory animals include those familiar species, the small rodents, which are specially bred for laboratory use and for no other purpose. These include mice, rats, guinea-pigs and hamsters, which together may account for over 95 per cent of all mammals used in the laboratory. Of the remaining 5 per cent, perhaps half are rabbits. These laboratory familiars are produced either by the laboratory using them or by another laboratory; by specialized breeding centres; or by commercial breeders.

The various sources of laboratory animals will be considered in the following section.

II. Collection

Long before there was any idea of breeding laboratory animals, experiments were devised and carried out on animals that had been collected, their species being chosen, no doubt, entirely on account of easy availability.

Today ready availability still to a large extent governs the choice of certain species which cannot be bred in captivity, or only with unreasonable difficulty.

A. COLLECTION FROM THE WILD

As has already been mentioned, collecting animals from wild natural populations raises a question of conservation. Species whose numbers in their natural environment are diminishing dangerously, and which are thus threatened with the possibility of extinction, are obviously not eligible for use as laboratory animals. Such include orang-utans, gorillas, and other species listed by the International Union for the Conservation of Nature.

Despite this, there have been reports of the illegal traffic in such protected species and even their use in damaging experiments for which less valuable species could have been substituted. The laboratory worker has an obligation, not only to refuse to use species in danger of extinction for damaging experiments, but to exert any influence he may have to condemn their use elsewhere.

Many experimental biologists are hampered by lack of an experimental animal that is entirely suitable to their work. Reproductive physiologists may be interested in the phenomena of prolonged gestation, delayed implantation of ova or storage of sperm, and for such work hystricomorph rodents and some species of insectivores may be invaluable. Others may want to work on monovular quadruplets and find in the nine-banded armadillo the only suitable subject. Physiologists may be concerned with water balance, renal concentration of electrolytes, and heat regulation; for them some of the desert rodents, like the sand rat or the rock hyrax, will be indispensable. For the preparation of scrub-typhus vaccine only the cotton rat will serve. For studies on environmental pollution by insecticides, concentration of residues at the far end of the food chain points to certain sea birds, to raptors, to bats or predatory carnivores as the laboratory animals of choice. For much biological teaching, and for some pregnancy diagnosis tests, species of frogs and toads have been used, often in large numbers.

All these animals, and many more, have been collected from the wild, and although some of them, such as the cotton rats and a few of the desert rodents, can be conveniently bred in captivity, many cannot, and the natural reservoir has to be tapped repeatedly. But with few exceptions the numbers taken are small, and represent tolerable cropping only.

Only a few species are demanded in such numbers that their collection is a threat to their survival, and in most if not all such cases there is another factor that constitutes a greater threat than collection for laboratory use. Thus, wild populations of frogs have decreased dangerously in many parts of the country, and are now not even seen where they used to be abundant. But schoolboys and others have been harvesting frogs for generations. Only in recent years have flies and other prey of frogs become less numerous, and those that are eaten often contain toxic levels of insecticide residues.

The prodigious demand for monkeys, above all for the rhesus monkey (*Macaca mulatta*) from India, for the preparation and testing of polio-myelitis vaccine over the last 15 or 20 years has led to a substantial export business from India in which many hundreds of thousands of monkeys have been shipped out each year. But for every one shipped, several others have died as a direct consequence of the hunt to catch them, in transit to the concentration areas, or as a result of illness after capture. Despite the enormous population of rhesus monkeys that India used to support, and even accord sacred status to, there are signs that this species has been seriously damaged. No doubt this is a relief to those whose crops were periodically raided by this most destructive animal, but its eventual total disappearance would be a tragedy, for which man alone would be to blame.

Similar considerations apply to many other species. Science would not be worthy of its good name if it took no account of the need for conservation of threatened species, or if it contributed to that threat, indifferent to the consequences.

B. STRAY DOGS AND CATS

Much of the controversy over the years about the morality of animal experimentation has centred on the use of stray dogs and cats that abound in many cities and have to be destroyed. Pet owners have been led to imagine that their favourite will sooner or later end up on "the vivisector's table", and it has not been difficult to work up this theme into a melodrama.

On the one hand there is, and has been for years, a need for dogs and cats for teaching in medical schools, for experiment and dissection in veterinary schools, for research in physiology, pharmacology and other experimental disciplines, for the study of diseases, especially of these two species, and for other purposes. On the other hand there is the vast number of strays that are regularly destroyed in any big city where control is organized. The number of dogs and cats destroyed is many times the number needed by the scientists: why then, they ask, can they not have the use of such strays as they need for their work?

In many cities, almost certainly a majority throughout the world, stray dogs and cats are available for laboratory uses, often with stringent conditions attached to that use to protect the animals from cruelty. But in Great Britain and in certain other states and countries, stray dogs and cats may not be used once they have been collected by whatever is the official collecting agency. They must, often by force of law, be kept for a certain time to be claimed, and failing that be destroyed, and they may not be offered for laboratory use.

This has led in the past to unofficial agencies anticipating the official ones in collecting strays and offering them to laboratories. Often enough the unofficial agencies have been perfectly legitimate business operations, conducted as openly and as honestly as any other business, and they have performed a real service to laboratories without endangering a single genuine lost pet animal. But, unfortunately, the competition between official agencies, such as the animal welfare societies and the police on the one hand, and the collector for sale to laboratories on the other, for the available strays has tempted the former into intemperate opposition to the legitimate needs of the scientist, and the latter into over-enthusiastic collecting. Pet animals have gone astray, though not in the large numbers that the anti-scientist lobby would like to make out.

The situation is unhappy. All attempts in Great Britain to reconcile the proper interests of the scientists and the animal welfare workers, even under the exigencies of war, have failed hopelessly. In some other countries understanding has been reached, and strays are made available, subject to reasonable safeguards. A few practical rules may be proposed for those who seek to supply the laboratory with collected dogs and cats, or to use them.

1. Make sure the animal is a genuine unwanted stray, look for a tattoo mark in the ear or elsewhere, a collar or other sign of identification. If one is found, the owner can be traced, and should be. A well-bred dog or cat is unlikely to be a stray, whether or not it carries a means of identification.

2. Make sure that the supplier is, as far as can reasonably be verified, an honest collector of strays, and not at the end of a chain of dog or cat stealing.

3. Keep a record of every dog or cat obtained, the sex, breed, approximate age, general description, state of health and source whence it came; and of its disposal.

4. Treat sympathetically any genuine enquiries from those who have lost pet animals and think they may have been stolen and sold to a laboratory. Most pet owners are reasonable people and they may be very distressed by their loss.

5. If an official pound for stray dogs and cats exists and will provide animals to laboratories, be careful to observe any conditions imposed. The use of stray animals for experiment is a delicate issue of public relations in almost all parts of the world.

C. FARM ANIMALS

The number of cattle, sheep, pigs, goats and horses used as laboratory animals is only a very small percentage of the total animals so used. For the most part they are too expensive or too difficult to handle in the laboratory. They are mainly to be found in veterinary institutes and schools, and in agricultural research centres.

All farm animals are, of course, specially bred, as animals ultimately destined for food or work. But they are not specially bred as laboratory animals, and so it is legitimate to consider them, when obtained for laboratory use, as collected; collected, that, is, from a source that exists quite independently of their usefulness as experimental subjects.

The scientist wanting to use farm animals in the laboratory will, therefore, obtain them as a rule from the market, and if he is wise he will seek the advice of someone who knows about farm practices and the best way of buying beasts.

For most laboratory purposes the farm animal used will in the course of its use be destroyed, or at least rendered unfit to go back to the market as a farm animal. But occasionally the nature of the experiment will be such that the animal is in no way disqualified from fulfilling its original role: perhaps some innocuous feeding experiments, or simple injections of evanescent effect, are all that the animal has undergone, and there is no practical reason why it should not be sold as normal, and recoup at least some of its cost. In such cases it is well to remember that there are various regulations to be complied with before an animal may be offered for slaughter. Application to the nearest office of the Ministry of Agriculture, or its equivalent body, should be made for information.

D. POULTRY AND RABBITS

Poultry, though smaller in size than the farm animals just discussed, are in much the same sense farm animals. Although they are used in considerable numbers in the laboratory, the poultry industry, which is very large, would exist if no birds were used there.

For many laboratory purposes, chickens obtained from reputable commercial sources will serve perfectly. The same is true of embryonated eggs and day-old chicks, which are obtainable at any time of the year from commercial hatcheries. Whatever numbers are needed by the laboratory are only a very small fraction of those being produced for the poultry industry. It is only when there are special requirements, such as certain freedom from specified viruses, that commercial sources are unlikely to be able to supply what is wanted.

There is also a rabbit industry in Great Britain, as in many other countries, but of a size many times smaller and much less sophisticated than the poultry industry. In addition, there is a lot of backyard breeding of rabbits, especially in country districts. Overall, the great majority of rabbits bred are for meat, and a substantial number for fur as well; but a minority are produced for laboratory use, and many of them come from small breeders who are more interested in the laboratory market than in the meat and fur trades. There is a general tendency, however, for the very small breeder to be losing the laboratory market, which is going to the larger, more intensive breeder.

Nevertheless, a laboratory rabbit breeder will always have an outlet, if a much less profitable one, for any surplus animals he may have. At the same time, the large intensive commercial rabbit breeder will find little difficulty in meeting laboratory demands out of his regular production, the bulk of which is for meat and fur. Thus, as in the case of farm animals and poultry, the laboratories are in effect collecting their rabbits from a source that would exist even if there was no laboratory demand.

E. PROS AND CONS OF COLLECTION

Collection can, therefore, be from wild populations (most primates, many species that are not commonly used in laboratories); from quasi-wild populations (stray dogs and cats); from branches of the animal industry (farm animals, poultry, rabbits); and also from the fancy and pet traders (guinea-pigs, mice, hamsters, rabbits, budgerigars).

In every instance the production of the animals is mainly for a purpose other than laboratory use, and therefore standards of quality are not related very closely to laboratory needs. Some attention may well have to be paid to the health of the animals after they come to the laboratory, especially that of primates, dogs and cats, and this is not only costly but may introduce a hazard to the health of the laboratory staff. There have been several deaths among laboratory staff from virus infections contracted from monkeys; helminthiasis from dogs, leptospirosis from rodents, salmonellosis from a variety of animals, fungal infections, and a number of other conditions, have all been reported.

On the other hand, some species breed little or not at all in captivity, and can only be obtained in sufficient numbers by collection. This applies to almost all primates today, to many less common laboratory species, to frogs and toads, and in the view of many to dogs and cats (but this will be discussed in the next section). Collection is unavoidable. It is also, even for many species that can be bred easily in captivity, much cheaper, because the natural environment has supported the production and growth of the animals, which in captivity would have been a charge on the breeder.

The following is a summary of the pros and cons of collection of animals for laboratory use:

Pros

1. They can be obtained as needed from the natural reservoir or population.

2. The cost of providing them only starts when they are collected: before then they are a charge on their natural habitat.

3. A large natural wild population can offer a choice of age and size.

4. Collected dogs and cats are a fraction of the cost of bred animals.

Cons

1. Attention must be paid to the claims of conservation of species threatened with the possibility of extinction.

2. The collected animal is almost always in need of treatment for infection and disease.

3. Animals collected from the wild are necessarily of unknown age, and have a low degree of uniformity (the same does not apply to farm animals, including poultry).

4. Collected dogs and cats offer temptations to the dealer to steal pet animals.

III. Special Breeding

Virtually all rats, mice, hamsters and guinea-pigs and many rabbits, as well as several other species not used in large numbers, are bred specially for laboratory use. Some laboratories breed all their own requirements, and may from time to time give or sell surplus animals to other laboratories. A great many specially bred animals come from breeding centres attached to universities, research institutes or under official sponsorship; such centres aim to provide animals to laboratories for experimental use, or perhaps only breeding nuclei of special strains, and they are likely to charge an economic price for them. Lastly, there are commercial breeding centres, which aim to supply animals for sale, in the ordinary way of trade.

These three main sources of specially bred animals will be considered separately: namely, the user-breeder, who breeds all his animals, usually in the same place as that in which he uses them; the breeding centre, usually but not always officially sponsored, and serving a number of laboratories, such as a whole university or a pharmaceutical complex, and sometimes outside users as well; and the commercial breeder, who breeds for profit and sells to any user. In what follows, the terms will be used with these precise meanings.

A. USER-BREEDERS

The first group of sources to be considered is the user-breeder, who breeds animals primarily for his own use, but may from time to time have a surplus to give or sell to other laboratories.

There are several reasons why a laboratory may decide that it should breed its own animals. The nature of the work may demand it; for example, in genetical research, breeding is part of the research and cannot usually be divorced from it. There may be a need for control of quality in breeding, which cannot be met satisfactorily elsewhere; or the user laboratory possesses a particular strain of animal which is essential for its work, and is reluctant to allow the strain to be supplied from another source. Alternative sources outside the experimental laboratory may not exist, or may be too far away to offer a satisfactory supply of animals. Facilities for breeding may exist in the laboratory, which it may be thought wasteful not to utilize. There may be staff who have been accustomed to breed the animals that are needed, and who are anxious that they continue to be able to do so.

Whatever the reasons, they should always be examined rather critically, because some of them may be less convincing than at first sight appears.

A laboratory doing its own breeding will have on the whole a smaller demand for any one species or class of animals than a breeding centre supplying a number of laboratories. It will thus find it more difficult to supply large numbers of animals of a given size and age or weight on a given date.

By the same token, there is likely to be a higher proportion of wasted animals that fail to fit the demands of the investigators, or of frustration on the part of the investigator who cannot get all the animals he wants when he wants them. This is a penalty of smaller-scale breeding, but it is often partially or completely overcome by close liaison between breeder and user, who work to meet each other's difficulties and arrive at a practical compromise. The frequency with which this happens perforce in circumstances where the user laboratory is doing its own breeding, and the (usually) small penalty in terms of inconvenience to the user, suggest that there is room for improvement in the relationship between laboratory and outside breeders, where an outside source is used. But this will be discussed in the next section.

B. BREEDING CENTRES

Breeding laboratory animals is a skilled technical operation. When it comes to breeding them in large numbers, with due regard to economy, additional skills are called for. So it has come about that in many places breeding centres have been established where these skills can be concentrated and placed under professional scientific direction.

Such breeding centres may supply animals for experimental use, at an economic price, to all laboratories asking for them; or they may serve only those laboratories that are controlled by the same organization that has established the breeding centre. A government research council or scientific department may decide to set up a breeding centre to serve the needs of all its constituent laboratories, but not particularly associated with any of them. Alternatively, there may be one or more officially sponsored national breeding centres.

In another type of breeding centre animals in numbers needed for experimental use are seldom supplied, but a collection of special strains is kept, breeding nuclei of which are regularly available for expansion in the user laboratory or elsewhere. Such centres in Europe are the Laboratory Animals Centre of the Medical Research Council in England; and the Centre de Sélection et d'Élevage des Animaux de Laboratoire in France. An official centre which performs both this function and that of supplying substantial numbers of experimental animals is the Zentralinstitut für Versuchstierzucht of the Deutsche Forschungsgemeinschaft in Germany. All these centres are national in character. Similar ones exist in other countries: in the Netherlands, in Czechoslovakia, in Hungary, in USSR, in Norway and in Sweden (see Chapter 11). Breeding centres that are sponsored by sectional interests, such as universities, research institutes or pharmaceutical companies, exist in many places. One of the largest in Europe is in Switzerland, set up by a consortium of pharmaceutical companies.

The breeding centre established by those interests that need to use the animals produced there should, of course, be run economically, but this may not be an absolute condition for its existence. Presumably the decision to establish such a centre in the first place was taken because the animals it was designed to produce were not available in the right numbers and of adequate quality elsewhere. Since the first purchase cost of the experimental animal represents only a small fraction of the cost of the work for which it is used, the most economical way of organizing and running a breeding centre may not be as important as that it should produce the animals. To get the right animals in the right numbers at the time they are needed may well be much more important than cutting off a small margin in their cost of production. Thus it must not be expected that economic operation is the first consideration of such breeding centres, and in fact they often do not operate with notable economy; yet they can still be justified by their product, even if the price is higher than it could be. If, however, the product is not in every way satisfactory, the breeding centre may well be regarded as a profitless extravagance.

A further function of breeding centres, especially those of official rather than industrial sponsorship, is research. It is often forgotten that a large breeding colony offers opportunities for population studies that cannot be found elsewhere. Such studies are compatible in many ways with the functional operation of a breeding colony. For example, a study of the spread of infection, not necessarily pathogenic, in a large colony can be undertaken without the deliberate introduction of a specific microorganism, because in almost any colony the acquisition of new microorganisms is likely to take place in any event from time to time (see Chapter 5). Physiological studies on conditions of caging, and on the physical and social environments of the animals, are often not possible in small colonies but are relatively easy to conduct in large ones. Methods of breeding, and certain reproductive observations, can be studied also under large-scale conditions but not in small colonies.

A breeding centre, therefore, has three main functions. First, it is a source of animals for experimental use. Second, it is a source of breeding nuclei of defined genetic or health status, to be used for expansion elsewhere. Third, it is in its own right a centre of "research on methods of research" (Medawar, 1957); that is, of research into the nature of laboratory animals and their use. If it fulfils these functions reasonably well, the fact that its animals cost more to produce than if it were purely commercial is of little or no importance.

C. COMMERCIAL BREEDERS

For many years there have been individuals or organizations who have seen in the supply of animals to laboratories a means of livelihood. Some of these suppliers breed all or almost all the animals they supply in commercial animal farms; others, although they may do some primary breeding, obtain a varying proportion of their animals from others who are probably primary breeders but in a smaller way of business. The primary breeder is properly called a breeder, while the one who buys from others to re-sell to the laboratories should be designated a dealer.

The dealer has a bad name because too often in the past he has been none too scrupulous about the sources of the animals he buys, or about their quality. It is true that he can often offer a very good service to the laboratories, in terms of obtaining the right number of animals at short notice; but unless he has control over his sources the animals supplied to him and that he passes on to the laboratories will be a very mixed collection, both genetically and from the point of view of infection. With certain limited exceptions, the dealer as a source of satisfactory laboratory animals is of little serious interest.

The exceptions concern the more costly laboratory animals, such as dogs, cats and above all primates. Quite apart from the problem of matching supply and demand, which will be discussed in the next section of this chapter, a commercial breeder of, say, dogs may find it prudent not to have all his breeding stock in one place, for fear of epidemic disease. He may provide a better service by establishing satellite breeders, under his general control, on which he can draw if any one of his sources is put out of action by disease or any other cause. But this is an exceptional case; similar arguments cannot apply to laboratory rodents or rabbits.

The commercial breeder, then, is a primary breeder, whose business it is to produce laboratory animals and to sell them to the laboratories. Since he is in business he has to make a profit and therefore, unlike the official or academic breeding centre, the commercial unit must pay great attention to economical and efficient operation. The commercial breeder must be economical to stay in business, and he must be efficient if he is to offer the laboratories animals of the quality that they require.

The question of quality is probably more misunderstood than any other aspect of laboratory animals. An animal is good if it serves well the purpose for which it is used; this is discussed in later chapters on quality. An animal with a strictly limited, even a gnotobiotic, microflora is certainly more expensive to produce than one less closely defined, but for many experimental purposes—toxicological screening, for example—it may be no better, or even inferior, and thus of lower quality.

For other purposes—for example, in some fields of immunological research—the more strictly defined animal is essential, and therefore must be available: animals less well defined are not just less useful, they are no use at all. It is therefore impossible to say that an animal with a given microflora is superior to another with a different microflora, just as it is impossible to say that red cars with four gears are better than blue cars with three.

It is against these considerations that commercial breeders have to be judged. In general the successful commercial breeders are relatively big, for only large-scale breeding can be organized economically. (Small backyard breeders, some of which may produce good general-purpose rabbits or guinea-pigs, are an apparent exception to this statement, but if all their labour and overheads were taken into account it would be found that their breeding is far from economical in comparison with the big breeders.)

Large-scale commercial breeding of mice and rats, and of some other species, has been established in USA for 30–40 years, and has increased greatly in the last 20 years. More recently it has developed in Great Britain (which had one or two important commercial breeders before 1940), and on the European Continent. Today, important commercial breeders can be found in England, France, Germany, Switzerland, Italy, Denmark, Sweden, Netherlands, Czechoslovakia, and some others of smaller size in most of the countries of Europe.

In 1950, the Laboratory Animals Centre, a unit of the Medical Research Council, introduced in Great Britain an accreditation scheme for commercial breeders, which had two main objects. The first was to investigate the extent to which commercial breeders were capable of meeting future needs, which even at that time were expected to become more and more exacting. The second object was to sort out and so far as was practicable control the best primary breeders by the granting of an accreditation certificate to those who were willing and able to maintain acceptable standards of husbandry and breeding practice.

In the years following the introduction of this scheme, there was a considerable increase in the number of animals used in laboratories, but the accredited commercial breeders enjoyed a lower share of this increased market than did laboratory breeding. It seemed that the rising demand for quality was not being matched by an increased ability on the part of the commercial breeder in Great Britain to produce good animals: he was losing the confidence of the laboratories (which he had never had in unqualified measure in any event).

Around 1960 some more far-seeing commercial breeders in Europe sought to follow an example that had been set in USA, in establishing breeding units under professional scientific control, in order to produce laboratory animals as good as those produced in the best breeding centres, and sell them at a price that would undercut the production costs of such centres. Between 1960 and 1970 a number of more or less sophisticated commercial breeding units were built in Europe, all of which were in charge of scientists and veterinarians, or had ready access to their skilled services. These commercial units were a far cry from the old backyard breeders, and the best of them did indeed produce animals—rats, mice, hamsters, guinea-pigs, rabbits, cats and dogs—of excellent quality and found a ready and profitable market for them. The biggest animal users were the pharmaceutical manufacturing companies, and most of these are severely cost conscious; they found that they could buy animals more cheaply and conveniently from good commercial breeders than they could produce them themselves. Some universities and research institutes also turned to the commercial breeders for at least a part of their needs, but there was less tendency for them to do so because universities are inclined to be less cost conscious and they often have a well-established tradition of doing their own breeding.

Economically, the rise of the good commercial breeder in Europe was very commendable, and deserved all the support it got. But there is another side to this. It has already been explained that economical breeding is only of secondary importance to the officially sponsored breeding centre, but it is of vital importance to the commercial breeder. In any commercial operation there are three claims on the business: that of the investors in the company, without whose investment the company would never exist; that of those employed, without whose work and skill there would be no product to sell; and that of the customer whose needs must be met, if business is to be done. In the commercial world there is sometimes a tendency to give undue prominence to the first of these three claims, to the detriment of the others.

A business that has established a reputation for excellence will carry on under its own momentum for some time after standards of work and of service have been allowed to fall; but not indefinitely. After a time the customer will complain and look elsewhere for his animals, and may be persuaded that the only solution is to breed them himself; and the commercial breeding unit, all commercial breeding units collectively, will suffer a decline in reputation.

Since the best commercial breeding units have actually set new standards in breeding efficiency and economical operation, such a development is to be greatly deplored. It would seem that in the field of laboratory animal production the commercial breeder of all the common laboratory species ought to have the advantage, and would have if his reputation generally were as high at that of the best; but until there is some way of knowing which commercial breeders are committed to giving as much weight to the claims of their staff and of the customer as they give to the strictly financial side of their operation (which is undeniably of vital importance also), laboratories may be excused for having some misgivings about entrusting their animal needs to the market place.

The accreditation scheme of the British Laboratory Animals Centre did attempt to guide users in their choice of commercial breeders, but its success, even after 20 years, has been partial only. The reason for this is probably to be found in the fact that only the biggest and most sophisticated commercial breeders are capable of understanding efficiently what is required of them, from the point of view of defining the quality of their animals, while an organization like the Laboratory Animals Centre, for all its special knowledge of standards of quality, has not the knowledge and experience of large-scale breeding that is an essential qualification of a commercial breeder. On the other hand the small commercial breeder, especially of animals other than rats and mice, has unquestionably had access through the accreditation scheme to technical help that might otherwise have been beyond his reach, and also to a market that he might otherwise have missed.

The most promising solution to the problem is for the best breeders to cooperate in setting realistic standards for their breeding operations, governed primarily by what they can know of the user's requirements; but for official bodies, such as the Laboratory Animals Centre, to collaborate with them in making independent assessments of the standards actually attained by the breeders. The initiative, in short, should come from the commercial breeders, but it should be guided if necessary, and endorsed if reasonable, by an independent authority.

Nowhere in the world today has this state yet been achieved, but there is nothing utopian about it. A parallel exists in the pharmaceutical industry, where close collaboration exists between individual manufacturers and the various official bodies whose task is to set standards for drugs. The relative safety of drugs that exists today could not come about except for the willing and wholehearted co-operation of the pharmaceutical industry, but the existence of official bodies setting standards of safety and efficacy relieves the industry of the unacceptable task of being judge of its own activities.

If the commercial production of laboratory animals is to gain the ethical and scientific recognition that it should have, it will have to live down the misgivings of the past, and set itself standards of quality and service that the user must have and that can be independently vouched for. This is a professional approach for the laboratory animal industry; it is the only one that is worthy of it.

The alternative is to rely on official breeding centres, or laboratories doing their own breeding. This is likely to be less economical and may also fail to provide the service as well as the quality that the user expects.

IV. Matching Supply and Demand

There are several species of laboratory animals, and reference has already been made to those that are specially bred for this purpose and no other, and so are not otherwise disposable.

All these animals come in two sexes, and in a wide range of weights and ages. Sexes do not change, but ages and weights increase daily, so that a population of laboratory animals is continually reclassifying itself according to these criteria.

These rather pedestrian considerations have a crucial effect on matching supply and demand. If too few rats are produced one week, the investigator will have to wait until next for his animals, and it can be both frustrating and expensive for him. If too many rats are produced this week, more than are needed, they cannot as a rule be kept until next week, because they will have grown out of the desired range of weight and age. They are therefore wasted; and although the wastage of a number of rats is less expensive, as a rule, than holding up work for lack of animals, it is still undesirable. An attempt must all the time be made to mitigate these disadvantages, by exercising control over production, by deciding at what point to kill surplus animals, and by encouraging trade and exchange.

A. CONTROL OF BREEDING

The times from mating to weaning of the young so produced are given in Table 2.1. These times are theoretical minima; in practice some allowance must be made for delayed matings or prolonged pregnancies, for infertility and for pre-weaning losses.

Table 2·1

Number of days from mating to weaning of progeny

	Mating to birth	Birth to weaning	Mating to weaning
Mice	20	18	38
Rats	21	21	42
Hamsters	16	21	37
Guinea-pigs	63	20	83
Rabbits	31	42	73
Cats	65	42	107
Dogs	60	56	116

The animals that are to be used today, therefore, are the product of a mating that took place some weeks ago, a period made up of the gestation period plus an allowance for delays, the period spent by the young in the nest from birth to weaning, and the interval from weaning until the age at which the animals are used. This sum may be called the production time.

Of course, few investigators are asked to place all their orders for animals so far ahead as this, but for special orders, such as very large numbers of one sex in a narrow age range, they may be compelled to. In general, a proportion of the animals produced by a breeder—of whatever kind, user-breeder, breeding centre or commercial—are committed to long-term orders, and only some are available at short notice, much shorter than the corresponding production time.

The breeder plans his breeding to take into account a fairly firm estimate of long-term or standing orders. In prudence he must plan to overbreed, in order that he may be sure of meeting these long-term orders, despite any vicissitudes that may affect his production, and he will rely on short-notice demands to take any surplus he has from week to week. In practice it is found that, for rats, mice, hamsters and guinea-pigs, and perhaps some other species as well, overbreeding by some 25 per cent is prudent; that is, the breeder will plan to produce some 125 per cent of his known long-term commitments. He may well produce more than this, in the belief that short-term demands will exceed the contingency of 25 per cent minimum that prudence demands.

The breeder will also seek to obtain from the user as accurate and as long-term forecasts of animal needs as he can. He will never be as successful at this as he would like to be, for investigators can seldom see exactly where their work is taking them, unless it be a routine type of testing that is entirely predictable. There will in most cases also be seasonal fluctuations in demand, some of which were discussed many years ago by Paterson (1953).

Armed with the users' forecasts, or his own estimation of them, together with a knowledge of seasonal fluctuations and a measure of intuition, the breeder can then plan his production. If he is skilled at his task, he will achieve a highly accurate degree of predictability of future production; more especially will this be possible if he adopts some of the methods of regulating production and standardizing growth that are referred to in Chapter 10. It is clear that his ultimate success in satisfying all demands without wasteful excess breeding will depend on all the factors involved: accurate forecasting of demand with short-term orders taking up the necessary excess breeding, highly predictable production in the required range of weights on given dates, and (unavoidably) a sound intuition.

B. DISPOSAL OF SURPLUS

Large-scale breeders of rats for a variety of users will normally plan to supply to laboratories not less than 60–70 per cent of all the rats they produce: the corresponding figure for mice is 80–90 per cent, and for guinea-pigs it should not be below 90 per cent. For rabbits the figure can be less, because surplus rabbits can be sold for meat and fur, although at a price well below that obtainable from laboratories. For most other specially bred species the percentage sold to laboratories should be virtually 100.

The wastage of rats (30–40 per cent of production) and mice (10–20 per cent) is considerable. Some of it will be accounted for by the animals not being in the right weight or age range when they are needed: some by the disparity of numbers of each sex demanded, male rats being often demanded in greater numbers than females. But some of the wastage is due to the need to overbreed in order to meet long-term orders, and to the failure of short-term orders to take up this necessary but often unpredictable surplus.

The surplus rats and mice cannot be kept for any length of time, because the demand for older animals rapidly diminishes. (Obsolescence affects guinea-pigs more slowly, and rabbits more slowly still.) There will be a need, therefore, in most weeks to kill some surplus animals. One of the more difficult decisions that the breeder has to take is when to kill his surplus. If some of his rats are not sold at 5 weeks of age, is he likely to have a demand for them at 6 weeks or older? and if so, how many? and is it worth his while keeping them on, to use space, labour and food that could perhaps be better employed in increasing his breeding stock? Only an intimate knowledge of the demand the breeder is serving will enable him to find the best answers to these questions.

Ultimately, surplus animals must be disposed of, to make room for others coming along. Rats, mice and hamsters, and also guinea-pigs, are usually killed, because there is no other use for them. Provided they are killed painlessly, there are no serious grounds for objection. But it should be remembered that those whose work has been to breed the animals should not be asked to kill them; it would be asking of most people too great a dichotomy of spirit, for the skilled breeder, although often willing to kill sick or injured animals, will usually resent having to destroy good animals that he has produced, just because there is no demand for them.

However, a surplus of dogs, and even cats, raises a more difficult problem. The destruction in cities of strays is often carried out by animal welfare societies, because they at least should be able to ensure that this is done painlessly. But a large dog breeder would find a formidable resistance, not only on the part of his staff, but from many other quarters, to the destruction of a substantial number of, say, 9-month-old beagles that he had bred but could not sell. If there was no market for them, why did he breed them? Moreover these animals are much more expensive to produce than small rodents, and to raise them to maturity only to kill them is financially unacceptable.

The problem of surplus, therefore, is not the same with dogs or cats as it is with rats and mice. They take many months to produce and grow, from mating to delivery to the laboratory, and the breeder must have a fairly accurate forecast of future demand.

There is a need here for collaboration between user and breeder, so that each may understand the other's problems. Animals are not like groceries, that can be stockpiled against need; they are produced, grow and pass through the age when they are useful all too quickly. If the user can, in the planning of his work, take this into account, he can help the breeder to avoid overbreeding and consequent wastage of animals; in return, he will receive from him a better service, and probably pay a lower price.

1. Methods of Killing

A booklet entitled *Humane Killing of Animals* is published by the University Federation for Animal Welfare (UFAW). In it will be found a choice of methods of killing various species, all of which are, when properly carried out, painless or almost so.

However, if the need arises to kill several hundred rats or mice that are not wanted for experimental use, most of the methods described in the UFAW booklet will be found to be not very practical.

For the humane killing of large numbers of laboratory animals, a gas chamber is necessary. This is an airtight box of suitable size, in which a number of cages may be placed. The box is then closed and the air displaced by a gas.

Carbon dioxide is a gas commonly used, and it has the effect of inducing anaesthesia before the animal actually dies. But the concentration of carbon dioxide has to be quite high in order to kill, and in a large chamber containing a number of cages the problem of replacing enough of the air with carbon dioxide is not easily solved. Killing by this method in such circumstances can, therefore, be slow and even uncertain.

Nitrogen is another gas that can be used for killing, and in sufficient concentration it is quick and humane. But the same problem of displacing enough air by gas exists as with carbon dioxide.

Neither carbon dioxide nor nitrogen is toxic or dangerous to handle in the ordinary way, but when they are used in large gas chambers for killing, there may be enough gas to be a hazard to the operator, especially if the chamber is inside a building. In particular, carbon dioxide is a heavy gas and will flow towards the floor of the room: an operator who falls to the floor would therefore be likely to become asphyxiated very rapidly.

More practical than either carbon dioxide or nitrogen is carbon monoxide. This is highly toxic gas which kills rapidly and painlessly in low concentration. In a gas chamber there is no need to displace much of the air; a small amount of carbon monoxide will rapidly diffuse through the chamber, and in well under 10 minutes all the animals will be dead.

Carbon monoxide, being toxic, is dangerous to handle. The gas chamber must be outside a building, in the yard or on a windy balcony or roof, and operators must be trained to understand the hazards. An oxygen resuscitation kit must always be readily available in case of accidents.

The most convenient source of the gas is a cylinder of carbon monoxide, of the least expensive grade. The exhaust gas from petrol engines contains a proportion of carbon monoxide, but there are also present a number of other gases, some of which are very irritant and may cause the animals great distress before they die. Coal gas also contains a proportion of carbon monoxide (unlike natural gas, which contains none), but the other gases present are irritant, and a considerable degree of air displacement is necessary to obtain a high enough concentration of carbon monoxide; this introduces a risk of fire or explosion. Neither exhaust gas nor coal gas therefore is suitable for the humane killing of large numbers of surplus laboratory animals.

C. EXCHANGE AND TRADE

The depressing need, from time to time, to kill surplus laboratory animals draws attention to the possibility that while one laboratory or breeder is killing a surplus another may be in need of just those animals. More frequently, one breeder may be unable to meet a short-notice demand, to the inconvenience of the user; but another breeder of an equally suitable strain of the same species may be able to supply.

In this sort of situation there is clearly a need for some exchange of information between breeders of laboratory animals. Such exchange often does exist in the case of large user-breeders and also of breeding centres, but in the case of commercial breeders this is not always so. It is true that the British Laboratory Animals Centre publishes a twice-monthly list of surplus animals (see next section), but for animals such as rats and mice with a rapid obsolescence this service is much too slow.

What is needed is an understanding between breeders that may be compared with that which exists, for example, between competing airlines, or medical practitioners. An airline will transfer a reservation from its own plane to that of another line, to the great convenience of the passenger; and because this is the rule, the airlines gain at least as much as they lose. A single-handed doctor on his day off will have an arrangement with one or more neighbouring doctors to look after any emergencies arising during his absence, with full confidence that his neighbour will refuse to allow patients to transfer to him as a consequence of such help. This sort of confidence, this ethical principle, does not yet exist to any extent between commercial breeders, but it should; for if they were genuinely motivated, not only by the need for their own commercial success but also by the desire to give a good service, they would understand that an agreement to give and take, without poaching or fear of poaching, would be of great help to the user and of no disadvantage to them; on the contrary, it would be of distinct advantage.

V. Accreditation Schemes

Accreditation schemes for animal breeding have been in existence in many countries for a number of years. The principle guiding them is that a suitable authority, often a government agency, lays down certain rules and standards that are designed to ensure at least a minimum satisfactory level of quality for the animals being produced. Those breeders who accept these rules and whose animals meet the standards set are listed as accredited. The possession of a certificate of accreditation is then some indication of quality.

A. LABORATORY ANIMALS CENTRE ACCREDITATION SCHEME

In the more advanced countries of the world there are accreditation schemes of one kind or another for many species of farm animals, including poultry. The first accreditation scheme for laboratory animals was introduced in 1950 by the Laboratory Animals Bureau (now the Laboratory Animals Centre) of the Medical Research Council, in Great Britain. It originally covered only three species, namely mice, guinea-pigs and rabbits (see Lane-Petter, 1953, and for a recent review of the scheme, Townsend, 1969).

Today, after 20 years of operation, the scheme covers a great number of species, and has undergone certain other developments.

No accreditation scheme can be a guarantee of quality. Whatever tests are carried out in a breeding colony today will only show the state of the colony today: tomorrow may be different. One is dealing with probabilities only. If today's reports show that, in a sample examination, it is probable that a certain colony is free of a number of specific infections, and that the standard of husbandry makes it unlikely that such infections will easily gain access to the colony, it is probable that today's report may be equally valid tomorrow. But, as is explained in Chapter 5, it is much less likely to be true in 6 months' or a year's time. The standards of accreditation can therefore only be used as a guide, not a guarantee.

The Laboratory Animals Centre accreditation scheme has proved of considerable use in helping small commercial breeders to produce better animals. The larger breeders have not benefited much in this way, because a larger breeder, to be large, must already possess a store of practical and technical knowledge that any central agency will find it hard to surpass. The usefulness of the scheme lies in the listing of breeders, especially the smaller breeders, whose standards are considered to be reasonably good, and this list or register is freely available to users.

In addition the Laboratory Animals Centre receives from breeders information about surplus animals available to laboratories. This information is published twice monthly in a sheet called *Parade State*, also freely distributed to users.

These two aspects of the accreditation scheme, namely listing reliable breeders and publishing *Parade State*, are undoubtedly useful, and help to avoid the situation where a laboratory needs animals that a breeder has in excess and is destroying. There is another side of the scheme that appears not so practical.

In the account of the scheme given by Townsend (1969) a description is given of an attempt to place accredited breeders' animals in one of five categories, according to the nature of the microflora (and certain other observations) found in them. Lane-Petter (1969) criticized this category scheme, and regarded it as unworkable and perhaps not even very useful if it did work; but the progress of time will eventually cast more light on a very difficult question. At the back of this conflict of outlook are some fundamental problems, both scientific and philosophical, to throw light on which is one of the objects of this book.

B. OTHER SCHEMES

The Institute of Laboratory Animal Resources in the USA has introduced a number of standards for breeding many species of laboratory animals, and these have been found useful in other countries. Indeed, the allocation of government grants is conditional on these standards being observed by suppliers of animals. There are also several other standards, of care of animals, of training of animal technicians, and so on, and full information about them may be obtained from the Institute of Laboratory Animal Resources, 2101 Constitution Avenue, Washington, DC 20418, USA.

CHAPTER 3

Trends

I. Introduction

The laboratory animal is to the biological research worker as the chemical reagent is to the research chemist. The development of knowledge in both cases depends on the tools available and as such knowledge advances so does the requirement for more sophisticated tools. Advances in biological knowledge have proceeded at a rapid rate during the last half century and at a breakneck pace since the end of the Second World War. These advances have had a profound effect on the nature of the test animal required, the numbers utilized and the species selected. Fundamental investigation into the basic processes of life, down to the level of chemical function of intra-cellular structures, has led to further effort in related fields and to advances in medical research; developments in the chemotherapy of disease have led to the establishment of a large pharmaceutical industry based on intensive research activity; the expanding use of sources of ionizing radiations in industry and medicine has led to extensive research into their biological effects; the need to feed an ever-growing world population and a sophistica-tion of the living environment for man have necessitated the adulteration of food by artificial substances, requiring extensive biological testing in animals, to enable it to be stored for long periods and to maintain its palatability, acceptability and freedom from detrimental organisms. These are some of the fields of research which are responsible for the increased demand for laboratory animals. The nature of this trend in terms of numbers, quality and species will be examined in this chapter.

II. The Increase in Demand

In Great Britain the total of vertebrate animals used for experimental purposes can be accurately evaluated on an annual basis due to the requirements of the Cruelty to Animals Act, 1876. This Act requires that experimentation on living vertebrates is undertaken only by those licensed under the Act and that such licensees provide the Home Office with an annual return of the numbers used. These returns go back several decades, but for present purposes it is only necessary to examine the annual totals since 1935 since the numbers used prior to the war years are insignificant in comparison to present-day usage.

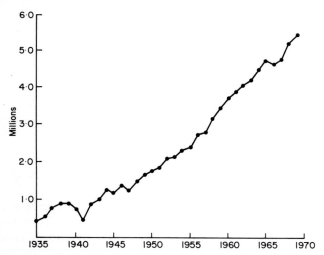

Fig. 3.1. Usage of vertebrate animals for experimental purposes in the United Kingdom, 1935–69.

The increase in the use of animals in Great Britain as shown by Fig. 3.1 is probably representative of usage in other parts of the world where biological research is undertaken on a broad front. It is not only the steady rise which should be noted but also the increase in the rate of this rise. If the use of animals was capricious and not closely integrated with the conduct of biological research, greater fluctuations in amplitude and frequency would be revealed. In fact, animal experimentation is an integral part of all facets of biological research and therefore the extent of usage is a direct reflection of the level of activity. The amount of money available for research has increased out of all proportion to the situation before the war, particularly in the medical, veterinary and agricultural fields, and this is directly reflected by the increase in animal usage.

The extent to which research within the biological field has increased over the last decade can be illustrated by a further analysis of the Home Office returns for the years 1959 to 1969. This is shown in Table 3.1. The figures under "Diagnosis" include use for purposes of public health or for the diagnosis of disease in man and animals; those under "Mandatory Assay" include the standardization of sera, vaccines or drugs required under the Therapeutic Substances Act, 1956, and Diseases of Animals Act, 1950,

Table 3.1

Analysis of returns to the Home Office, 1959–69

	Mandatory assay	Diagnosis	Cancer research	Total animals used overall
1969	1,453,939	277,641	337,207	5,418,929
1968	1,204,860	225,677	306,652	5,212,215
1967	1,238,294	250,884	315,319	4,755,680
1966	1,182,365	297,720	326,521	4,615,023
1965	1,331,288	259,992	358,310	4,751,060
1964	1,411,872	244,432	343,902	4,494,931
1963	1,173,535	318,851	269,658	4,196,566
1962	1,214,276	202,528	258,780	4,041,944
1961	1,113,874	244,409	233,273	3,896,581
1960	1,236,585	218,931	229,751	3,701,187
1959	1,154,118	203,856	276,263	3,493,022

and the requirements of official bodies in foreign countries in the case of exported medicines. The number of animals used in these categories can be seen to have altered little during the past decade while the overall usage of animals has increased by some 59 per cent. The proportion of usage for these "non-research" functions was about 39 per cent of the total in 1959 but declined to 32 per cent in 1969. An extrapolation of this trend will provide a short-term forecast, possibly extending for 10 years, of animal usage. The extended extrapolation beyond this period may bear little relation to actual usage since such variations as the choice of species, direction of research, funds available and quality required may drastically affect the numbers utilized. However, the short-term forecast is an essential guide in the laying of plans to ensure that biological research is not starved of its basic test material. The factors affecting long-term usage forecasts will be discussed in the course of subsequent sections.

This expected increase in quantity affects not only plans for the production of suitable animals, but also for the necessary maintenance and experimental facilities. A recent American survey (Report, 1970, a) indicates the
magnitude of the problem.

III. Species Trends

Of greater interest than an appreciation of the increase in the overall
numbers of animals used is an analysis of the trends in utilization of particular species. Such trends are not dictated by fashion but by the requirements to fulfil a scientific function. The dominant position of rats and mice
in the spectrum of species used has been maintained since the start of the
century—for reasons already outlined—and will probably remain for many
decades to come. The need to provide quantitative data, which overcome
variation in individual animals, is a prime requirement in most fields of
biological research. The small rodents fulfil this need very adequately and
will certainly continue to do so.

The publication of surveys of species utilization is an extremely rare
phenomenon and an assessment of trends must be made with certain
provisos in mind. The 1956 survey of usage in laboratories in the United
Kingdom, carried out under the auspices of the International Committee
on Laboratory Animals (Report, 1958), must be matched against a recent
survey carried out in the United States of America for the year 1969, under
the auspices of the Institute of Laboratory Animal Resources (Report, 1970,
b). This survey was compiled from data provided by a 70 per cent response
to questionnaires sent to 2290 laboratories. The comparisons are valid to a
large extent since both countries undertake extensive biological research in
universities, government institutes and the pharmaceutical industry.

A. RODENTS

Table 3.2 illustrates the main species groups utilized as revealed by these
two surveys. Rats and mice together show an increase in proportional
utilization, but the interesting feature is the increased popularity of rats at
the expense of mice. This trend may be due to a number of factors, two of
which are pre-eminent. Experimental screening and assay techniques, used
in pharmaceutical research, are becoming increasingly sophisticated and
demand a larger animal than the mouse to enable the surgical procedures
involved to be carried out; the second factor relates to the increasing
availability of high-quality commercially bred rats at economic prices.

It should be borne in mind that when considering trends and numbers
utilized, the relationship between the species, of initial and maintenance
costs and animal-house space required should be borne in mind. An increase

Table 3.2

Proportional usage of experimental vertebrate animals for the years
1956 (United Kingdom) and 1969 (United States of America)

	1956 survey UK	1969 survey USA
Mice	69·6	64·6
Rats	13·3	22·3
Guinea-pigs	7·7	1·7
Hamsters	0·3	1·6
Rabbits	1·5	1·1
Cats	0·3	0·3
Dogs	0·2	0·7
Primates	0·1	0·1
Birds	4·1	3·4
Amphibia	2·1	3·1
Other vertebrates	0·8	1·1
	100·0%	100·0%

in rat utilization of 9 per cent of total usage (Table 3.2) has considerable implications for planning and costing.

The trend towards greater utilization of rats can be expected to continue, since the increase in scientific knowledge produces a requirement for more detailed information for the furtherance of this knowledge; the size of the rat makes it more and more the animal of choice in medical research. The rat is also an ideal animal for the provision of quantitative toxicological data and its use in this field will increase as the demand rises for new drugs, food additives and pesticides.

The mouse is still the animal of choice in the field of cancer research, for reasons previously discussed. Although it is losing ground in other fields of research the overall quantity utilized is increasing with the expansion of research effort. Home Office returns demonstrate that in cancer research in Great Britain the numbers of animals used increased from 229,751 to 358,310 in the years from 1960 to 1965. The greater part of this expansion related to mice. However, while overall usage showed a steady increase during this period (Fig. 3.1), the increase in cancer research has halted since 1965. This may be due in part to the banking of transplantable tumour tissue using deep-freeze techniques, thus avoiding the necessity of maintaining tumours by continuous transplantation in mice.

The decline of the guinea-pig as an experimental animal was from 7·7 per cent in 1956 to 1·7 per cent in 1969, the latter figure being possibly true for Great Britain also, is probably a reflection of the rise in popularity of the rat where a rodent larger than a mouse is required. The latter has the advantages of faster breeding turnover, a simple dietary regimen and availability as a gnotobiotic animal. The use of guinea-pigs for the provision of material for *in vitro* pharmacological testing on isolated organs remains widespread, but the reduction in numbers used is also due to its decline as a test species in the diagnosis of tuberculosis, together with a reduction in the incidence of the disease itself. An interesting feature of the American survey is the apparent popularity of the hamster, whose usage figures almost approach those for the guinea-pig.

B. CATS AND DOGS

The Home Office returns under the 1876 Cruelty to Animals Act enumerate the usage of cats, dogs and *Equidae*. This enables trends in the use of these species to be followed with accuracy. The pattern of research is constantly changing and this has been particularly evident during the past 10 years where animal usage is concerned. The proportion of cats to the total of all vertebrate species used remained fairly constant between 1960 and 1969— 0·29 per cent and 0·27 per cent respectively. The same proportion for dogs for these 2 years were 0·20 and 0·32 per cent. The proportions used in 1969 in the United States were 0·3 per cent and 0·7 per cent for cats and dogs respectively. Although these figures appear low, the absolute numbers used increased considerably. Figure 3.2 illustrates this increase for the years from 1960 to 1969.

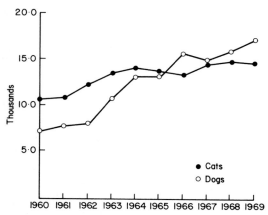

Fig. 3.2. Use of cats and dogs in Great Britain. 1960-9.

Increases for both species reflect to a large extent the requirements for experimentation in new surgical techniques, especially those concerned with organ transplantation and the need for larger animals which permit extensive response monitoring under extended periods of anaesthesia. The use of electronic equipment in such experimentation enables simultaneous multi-channel recordings to be made, providing a great width and depth of information to be obtained from a single animal. For this purpose cats and dogs are the species of choice. The additional increase in dog usage, when compared with cats, is certainly due to the requirement of official bodies for more extensive toxicological data on new biologically active compounds or food additives; for this purpose the dog is the universally chosen species providing supplementary data to that obtained from the rat.

The demand for these two species will undoubtedly increase. The establishment of the beagle as a laboratory species, and improvements in the techniques for breeding and holding cats, have done much to gain the confidence of the experimenter where previously he relied on strays of dubious origin and in a very poor state of health.

C. OTHER VERTEBRATE SPECIES

The use of non-rodent species has been shown to comprise about 10 per cent of the total. Certain among these species can be given special mention. Rabbits are often found in animal houses; although small in numbers their size dictates the occupation of a fair proportion of the available space. The 1956 Great Britain and 1969 United States surveys indicated rabbit usage of 1·5 per cent and 1·1 per cent respectively. It is unlikely that this represents a trend and it is probable that absolute numbers used have risen owing to the requirements for teratological data in this species.

Primate usage is given as 0·1 per cent in both surveys. Although the proportion is small the space required for these animals is considerable and the figure represents a high degree of importance given to this animal. The use of marmosets in pharmacological experimentation is increasing, and when more becomes known about the breeding, maintenance and physiological responses of these primates they may well come to hold an important position as an experimental animal.

The 1969 survey shows that amphibia were used to the extent of 3·1 per cent of the total. This figure may well relate to a special situation in the United States, since in Great Britain these animals are not now easily obtainable and are expensive. The usage of birds includes some 2·5 per cent for poultry in both surveys.

The usage of other vertebrate species is insignificant in comparison with the total and covers a vast range. Over fifty species are noted in the 1969 survey other than those groups already mentioned. The requirement for these species is in relation to specialized research objectives and is outside general considerations.

IV. The Increasing Demand for Known Quality

The experimental animal must fulfil certain basic criteria which will ensure that the results obtained from experimentation will be capable of direct interpretation and will not be misleading. The animal must suit the purpose for which it is required. This theme will be repeated in subsequent chapters as it applies specifically to considerations of environment, health, management and economics.

The main considerations applied to the choice of animal relate to its genetic constitution, its disease status and its nutritional condition. These will all be dealt with in detail in subsequent chapters. The history of laboratory animal usage is closely associated with an increasing awareness on the part of experimenters that the successful outcome of biological research is dependent on the nature of the animal used, with respect to the three criteria already mentioned. This does not necessarily imply that for a group of animals to be satisfactory they must have complete genetic uniformity, be free from all known pathogens and be nutritionally complete—although this last criterion is indispensable for most purposes.

✗ The chemist does not require an absolutely pure reagent for all his experimentation and similarly the biologist can utilize satisfactorily a laboratory animal which may not have a known genetic constitution or be free from all disease. The real trend in the selection of the test animal is the definition of quality in relation to purpose. ✗

The development of techniques in the field of germfree and gnotobiotic animals, and improvements in animal-house design and management, have made it possible for animals of a high quality to be available to those who are in need of such standards. The use of animals which are nutritionally deficient or which carry latent disease has led to costly failures in research and frustration in experimenters. Such failures result from variable and unreliable data provided by animals which do not reach the standard of health required for the purposes of the experiment, or from serious disease outbreaks which enforce the termination of an experiment before its completion.

The provision and maintenance of high-quality animals are expensive, but where the purpose requires such a standard it is economically unreasonable to be satisfied with less. That it is desirable in economic as well as scientific terms to utilize the advantages of a more closely defined animal is illustrated by the trend in most laboratories towards investment in more sophisticated accommodation, the production of disease-free animals and the improvement in quality of animals obtainable from commercial sources. The continuation of this trend towards the point at which only completely pathogen-free and genetically defined animals are utilized can be regarded as a swing of the pendulum too far in one direction. There will always be a demand for the less refined animal, since it is satisfactory for many purposes and cheaper to produce and maintain. For example, a cat which is free from respiratory disease and nutritional deficiency is a perfectly adequate laboratory animal for most purposes; to expend a great deal of effort in producing a completely defined animal would not make it any more acceptable.

In summary, the trend towards improved quality is progressing, since in many laboratories the facilities and animals are inadequate for the required purpose, but this trend progresses in conjunction with a growing awareness of the need to equate quality with experimental requirement.

V. The Need to Study Trends in Demand

The tables and graphs presented in this chapter illustrate the increasing demand for experimental animals. It is necessary to study and interpret such trends in order that research may not be deprived of animals which fulfil the requirements of quantity, species and quality. The examination of total usage trends reveals little of real value to those who have the responsibility for providing for the future. The increase in usage of cats and dogs is of far greater significance even if the numbers utilized are small in comparison with the total. The breeding and rearing of one cat may be equivalent in cost to the production of some 75 rats or 300 mice and the space required in the animal house for the experimental cat is almost as great. The increase in dog usage should be studied as a separate entity since the requirements for housing are specialized, elaborate and costly. The use of primates is increasing and may continue to do so; should provision be allowed for these species in an animal house under design? While individual laboratories have requirements which reflect their own particular research directions, a knowledge of trends in the use of species is of undoubted value to those who must plan facilities and who are responsible for providing the animals.

Surveys of animal usage in detail are too few and far between to provide data of a quality necessary to make intelligent extrapolations into the future. Not only must present data be accumulated but the views of senior research workers should be sought to provide a reasonably accurate forecast of future needs. The cost of animal accommodation, provision and maintenance is high, as will be seen in Chapter 9, but the cost of idle research due to the unavailability of suitable animals or the inadequacy of facilities is higher.

The use of small rodents as predominant experimental species for many decades has resulted in a wealth of relevant data relating to production and quality; variation in demand can be adjusted with reasonable speed because of their high breeding rate. The position regarding the larger species, cats and dogs in particular, is quite different. Not only has the demand for quantity increased but also the requirement for quality and specified breeds. Such trends demand additional effort in the study of suitable husbandry techniques and the establishment of an adequate number of breeding colonies to meet demand.

The subsequent chapters in this book will stress this need to study trends and ensure that future research requirements are met by an adequate supply of suitable animals.

VI. Substitutes for Laboratory Animals

An essential characteristic of man is the desire to enquire into the unknown; research into biological processes and the nature of disease has produced dramatic results in the improvement of his condition. The development of chemotherapeutic agents used in the treatment and prevention of disease is perhaps the most dramatic of advances made in the last two decades, but with the need to feed an ever-increasing world population considerable advances have been made in agriculture and food technology. These advances could not have occurred without the aid of the laboratory animal. There are those who consider that animals should not be used for such purposes and this point of view is in most cases presented with sincerity and deep conviction. These humanitarian considerations must be regarded in the light of a true perspective of the benefits accrued to mankind and the nature of the effect on the animal. In this latter context the Home Office returns for 1969 show that 88 per cent of experimentation involved only the procedures allowed under Certificate A—simple inoculation or administration of substances in food or water. Long- or short-term experimentation is only permissible in Great Britain provided the animal does not show signs of severe or enduring stress. If this occurs the experiment must be concluded by painless killing of the animal. Operations with subsequent recovery may only be undertaken under circumstances where, subsequent to the post-operative period, the animal shows no signs of severe or enduring distress.

The biological processes involved in the maintenance of life are complex and interrelated, and for most purposes biological research requires that these processes are all functioning during the course of experimentation. Experimental animals are extremely costly test material and where substitutes can be found these are preferred. It has been seen that some 32 per cent of laboratory animal usage is for the purposes of diagnosis and mandatory assay; in this field it could well be that future developments in chemical and biochemical analysis and *in vitro* techniques will reduce this usage to a negligible level. Although the experimental animal is a complete biological entity it is also a variable one; laboratory techniques can become sophisticated to the point at which the variability is far below that experienced in the living animal. Such techniques can be faster, cheaper and more definitive than assay in the laboratory animal. The use of microbial cultures in the diagnosis of tuberculosis is an example.

Chick embryos constitute a ready-made form of sterile whole animal culture. As such they have been extensively used in the culture of viruses. It has been suggested (DeBock and Peters, 1963) that the chick embryo could be utilized for the teratogenic testing of new drugs, since it has been shown to be sensitive to this effect of thalidomide. The ease of administration of the test substance, the availability of fertile hens' eggs and the economic advantages to be gained by the reduced maintenance costs are features which merit further investigation into the use of this medium in developmental studies.

The use of computerized mathematical models to replace living animals in biological research has been proposed. Since the basic physico-chemical reactions of living tissue are still largely unknown, it is difficult to envisage how such a system can replace the entire animal except in the very simplest of test situations. Because most research employing experimental animals has the human condition as the main objective, the use of animals such as invertebrates, as replacements for warm-blooded vertebrates, is hardly relevant since they are anatomically and physiologically out of context.

The use of tissue-culture techniques as a replacement for the laboratory animal presents many difficulties. Most research requirements depend on the presence of the whole spectrum of physiological interactions present in the vertebrate animal, and this cannot be reproduced in culture. Fundamental research into sub-cellular processes may be undertaken on cultures of vertebrate tissues or cell types. However, such cultures alter their nature in a short space of time, becoming moribund, reverting to a less differentiated type or even becoming malignant; in consequence, new cultures must be continually set up from fresh animal tissues—usually embryonic—supplied by laboratory animals of consistent health and genetic constitution. Research into processes occurring at the tissue level may be frustrated by difficulties of interpretation where the response is in an environment radically different from the *in vivo* condition.

While it is not too far beyond imagination to visualize toxicity testing in cultures of a wide variety of mammalian tissues, it is unlikely that any new substances would be permitted to be consumed by man without additional exhaustive testing in the whole animal—with negligible saving in the number of animals used. In view of the fact that the use of whole animals represents considerable financial outlay, that the animals themselves may fail the experiment for health reasons and that *in vitro* methods should be more reproducible and precise, the tendency in animal biological research will be to find substitutes for the laboratory animal. However, as far as the future can be foreseen there will always be a need to investigate response in the whole animal.

CHAPTER 4

Genetics

I. A Theoretical Colony

Consider a large population in which mating is in no way regulated. If it is in equilibrium, the probability is that any individual in the population will be homozygous in respect of about half the genes it carries, and heterozygous in respect of the other half, and the population may be described as 50 per cent homozygous (or heterozygous).

Note the conditions necessary for this state: a large population, free or random mating, and a state of equilibrium. Under laboratory circumstances these conditions will rarely if ever be met.

Such a population is, in fact, a theoretical figment, but it will be useful to discuss briefly the three conditions that, ideally, are needed for its existence.

A large population in this context means not only one comprising a large number of animals but also one in which the chances of reproductive contact between individuals throughout the population are unrestricted and indiscriminate. An individual of one sex thus has an equal chance of mating with any individual of the other sex, and the choice is coterminous with the population. In these circumstances, matings between closely related individuals will sometimes occur and this will lead to an increase in the degree of homozygosity over the theoretical starting value of 50 per cent. But provided that the population is sufficiently large, these closely related matings will be so rare as to be negligible. In a smaller population closely related matings will be less uncommon. They may have an important effect on the degree of homozygosity, and this can be calculated.

Random mating implies a statistically nil chance of any one mate being preferred to any other. This could theoretically happen if freedom of movement throughout the population was absolute, like the molecules in a gas, and if all individuals were equally nubile. In practice neither of these conditions is ever met. In large populations movement is not indefinitely free, and mating preference will be influenced by physical proximity. Uniform matability under conditions of free movement is also a theoretical state. In practice some individuals will show more reproductive energy than others, and thus upset the random picture.

In smaller populations movement may approach a degree of indefinite freedom, coterminous with the population, but the chances of related mating are increased. The matings can, of course, be controlled, according to a statistically random principle, or deliberately manipulated either to increase or to decrease the actual number of related matings, or in other ways.

In any event, departures from the theoretical state of random mating will constitute some degree of selection. If such a departure occurs without deliberate manipulation, it can be regarded as natural selection: if it is deliberately provoked, it becomes artificial selection.

II. Laboratory Populations

The theoretical colony discussed in the last section is not realizable, nor is it even desirable. Large populations are not susceptible to the degree of detailed and direct control that is necessary to ensure high quality. Actual matings will not be left to the laws of chance and the mating preference of the individual animals, and equilibrium will be constantly over the hill and possibly far away. What, in practice, is likely to happen?

Large populations are not susceptible to close control, because it takes a great deal of time and effort to examine animals in the exercise of such control. Many colonies are maintained for the sake of characteristics that are governed by their genetic composition, and this applies especially to inbred strains. If the desired genes are lost or altered, the value of the colony is diminished or lost.

Some of these characteristics may be as easily detectable as coat colour, but others may be specific responses to pharmacological procedures, specific susceptibilities or immunities, or small morphological features. The less obvious ones need to be tested for constantly, and their testing may entail much work. It is just not practicable to do this on a large scale, but if it is not done on a scale that bears a useful relation to the size of the colony, it has little value.

Thus, for the sake of ensuring that a given colony continues to possess at least those characteristics that make it useful, whether they be mutants, specific responses or any other genetically controlled features, a small colony is a necessity. Hence the first condition, a large population, can never be realized under conditions of direct control.

Control of the kind indicated implies that it is desired to maintain the wanted characteristics in the colony. To the extent that these are genetically controlled, this will mean selecting for them, first to fix them, and later to stop them becoming unfixed. In a small population some degree of related mating is inevitable and this will increase the degree of homozygosity of the population in calculable measure, even if no very close inbreeding, such as brother by sister mating, is resorted to for the sake of fixing a characteristic. Thus, the second condition goes overboard.

Lastly, the state of equilibrium, the third condition for a stable state of 50 per cent homozygosity, depends upon maintaining an environment of absolute constancy. This cannot be achieved in practice, so the third condition is also abandoned.

Laboratory animal colonies will always possess more than 50 per cent homozygosity. The smaller they are, the more closely inbred they are, and the more vigorously they are selected for certain characteristics, the more nearly will homozygosity approach 100 per cent. But that part of their genetic composition that becomes homozygous will not be the same in two colonies isolated from one another. Separate colonies are homozygous at different loci and thus differ from each other genetically, even though they have a common origin; and this applies even where they are not closely inbred.

For this reason, it is erroneous to suppose that a colony of, say, "Swiss mice" derived from a given source will, after many generations of isolated breeding, be virtually indistinguishable from another colony derived from the same source. The two colonies will be different, and the designation of each should indicate this. The practice of calling the daughter colonies by the same name as the parent colony, from which they have been separated by many generations, merely leads to the name losing any useful meaning, and it should be condemned.

III. Inbred Strains

For many purposes the user requires animals that are closely inbred. The accepted definition of an inbred strain has been published in respect of mice, and is generally accepted. It is as follows:

"A strain shall be regarded as inbred when it has been mated brother × sister (hereafter call b × s) for twenty or more consecutive generations. Parent × offspring matings may be substituted for b × s matings provided that, in the case of consecutive parent × offspring matings, the mating in each case is to the younger of the two parents." (Staats, 1968).

The effect of this degree of inbreeding is to increase the homozygosity to a maximum, which is virtually achieved within twenty generations. Theoretically, this means that for every gene the animals are homozygous. Falconer (1967) has shown by means of a useful diagram that the rate of increase of homozygosity as a result of inbreeding is rapid at first, but tails off as the value approaches 100 per cent. The achievement of 100 per cent homozygosity will be prevented or delayed by the appearance of mutations at any stage, and possibly by the nonviability of animals in the completely homozygous state, and these causes will tend to perpetuate the heterozygous state. Mutations are rare, but they do occur, and that is why, even after twenty generations of brother by sister mating, it is necessary to continue the process indefinitely. For when a mutation does occur, its isolation, detection and elimination are facilitated by inbreeding. On the other hand, the abandonment of inbreeding can hasten its spread throughout the population, especially if it is recessive or not disadvantageous, and this militates against, while not absolutely preventing, its manifestation.

If in a previously inbred population, inbreeding is discontinued, there will in time accumulate mutations, and the degree of homozygosity will decline—very slowly at first, but after several generations with quickening tempo. The decline is for all practical purposes negligible during the first ten or so generations, and this has a most important bearing on the large-scale production of authentic inbred strains. It has made possible the development of the traffic-light system of subcultivation of inbred strains, which is described in Chapter 10.

A. INBREEDING DECLINE

It is well known that increasing the homozygosity of a strain by inbreeding leads to a reduction in general vigour, and this will be reflected in such things as lower fertility, shorter life span, smaller size, slower growth, increased susceptibility to disease and, it should be noted, a greater lability of response to small changes in the environment. This reduction in vigour is called inbreeding decline, and it may partake of any or all of the above failings in greater or less degree.

The simplest way of demonstrating inbreeding decline is to introduce continuous brother by sister mating into a colony that has never previously been closely inbred. In the first generation or two little change may be noted, but by the third generation some or all of the failings characteristic of inbreeding will begin to appear, and by the fifth to seventh generation the odds against the strain surviving are long. In fact, the history of many established inbred strains in existence today is one of perseverance in the face of difficulties arising from inbreeding decline; the existing strains are the few survivors from a far greater number of attempts, nearly all of which failed.

However, if a strain can be brought to about the tenth generation of brother by sister mating, it will have approached very closely to its ultimate maximum of homozygosity. If it still survives, this must be because it possesses a set of genes sufficient for survival, and is homozygous for most or all of them. Further inbreeding will therefore be followed by little or no further decline.

The fact is that inbred strains generally are less vigorous, less fertile and altogether more difficult to breed than non-inbred, and no research worker would be likely to choose an inbred strain if an outbred one would do. Homozygosity itself, apart from the presence or absence of particular genes, is disadvantageous, and homozygous animals are, as laboratory animals, less robust and more difficult to maintain than heterozygous. This is unfortunate, for it represents a serious price to pay for the advantages homozygous strains have for some kinds of research, where genetic uniformity is a necessity.

The lack of robustness is a serious disadvantage, which certainly contributes to, if it is not entirely responsible for, a paradoxical lack of uniformity in inbred strains. The phenotype—what the animal actually is—is the result of the interaction between the genotype and the environment. Since the genotypes of all the individuals of a homozygous strain are identical, and the environment is apparently the same for all, it might be expected that the phenotype would also share in this uniformity. But in practice this is found to be not necessarily so. A great volume of work has been done on this subject, and it still continues, but it is sufficient here to point to two conclusions, about which little doubt remains. The environment can only be considered uniform so far as it affects the animals (and this is the sole criterion that is of interest), if it is not only truly uniform, but uniformly good. Any failure to provide an optimal environment will have an adverse effect on any animals but, owing to the reduced robustness of inbred strains, small shortcomings in the environment, which would have little or no manifest effect on heterozygous animals, may have exaggerated effects on homozygous ones. Whatever the precautions that are taken. even when all the animals of a colony are kept in the same room with conditions as nicely regulated as can be contrived, the micro-environment in different cages will not be exactly the same everywhere. The relatively unstable biological systems represented by homozygous animals will consequently react like over-sensitive instruments to these small variations in environment, and the expected uniformity will not be realized. The only way of overcoming this difficulty is either to stabilize the environment more carefully or to abandon the demand for homozygosity.

B. HYBRID VIGOUR

Granted that for some work genetic uniformity is a necessity, and that optimal uniformity of environment is a future hope rather than a present reality, is there a solution? There is, and a very simple one. If two inbred strains, preferably dissimilar, are crossed, the genetic contribution of either strain to the first filial (F_1) generation will always be the same, and so the members of the F_1 generation will be as genetically uniform as their parents (Fig. 4.1). But, in so far as the two parent strains were dissimilar, the F_1

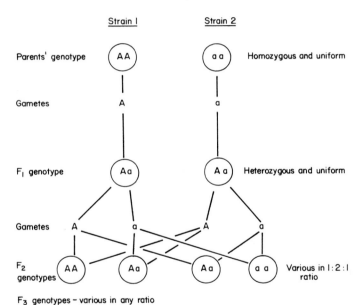

Fig. 4.1. When inbred strains are crossed, only the F_1 generation is genetically uniform, while at the same time being heterozygous. The figure shows what happens at a single locus when two strains, homozygous for different alleles, are crossed.

generation will be heterozygous, with all the benefits that the heterozygous state confers in the way of vigour. Thus, the desirable state is achieved of genetic uniformity—all the F_1 crosses must be as genetically identical with each other as were the members of the parent strain—combined with hybrid vigour. The production of F_1 crosses, or hybrids, between two inbred strains is thus a neat way out of a difficult dilemma.

This only applies to the F_1 generation. If these are mated together, being heterozygous they will undergo segregation and recombination of genes, and the next generation (F_2) will have lost its genetic uniformity. Thus, the production of genetically uniform heterozygotes is limited to the F_1 crosses between inbred strains. To produce them the inbred strains must still be maintained indefinitely.

Genetically speaking there are, then, three classes of animals from which to choose: outbred, which are more or less heterozygous and genotypically various; inbred, which are virtually homozygous and genotypically uniform; and F_1 crosses. The first group is vigorous, robust and stable but, being genotypically various, there will be some variation in phenotype. The second group is less vigorous and robust and, in the presence of even small irregularities in environment, labile, leading to phenotypic variability. The third group has it both ways, at least theoretically. It would be very satisfactory if it also turned out to be so in practice, and indeed it often does, but not always. Grüneberg (1955) in a masterly comparison of certain morphological characteristics in animals from the three groups, showed that sometimes one, sometimes another, exhibited the greatest uniformity of phenotype, and that the choice followed no rule. It was, in fact, unpredictable, and this, in the light of more recent work (not confined to morphological characteristics), remains the conclusion today. The research worker who seeks a high degree of uniformity in respect of a particular characteristic will therefore be obliged to make his choice only after a direct comparison of the three possibilities; there is no short cut by way of *a priori* assumption.

Cholnoky, Fischer and Józsa (1969) have proposed another kind of genetic manipulation, entailing the ordered crossing of inbred strains to produce maximum within-group variation, and they suggest that such populations may have a special application in pharmacological assays. They describe the groups of animals as mosaic populations, and present details of the mating systems by which they may be achieved.

C. EXPANSION OF INBRED STRAINS

From what has been said about the practical difficulties of controlling large colonies, it might be inferred that the production of large number of animals of an inbred strain is not possible. A further difficulty is the tendency for divergence to occur between two or more lines that are separated from a common ancestor by many generations. The evidence for this tendency is compelling; sublines of inbred strains that have been separated sooner or later come to differ, and this divergence may be expected to be recognizable after some ten to twenty generations. In maintaining an inbred colony, therefore, the aim must be to keep a single line in existence perpetually.

On the other hand, too narrow an adherence to this rule will neither produce a useful number of animals nor give a wide enough field for selection; and both a reasonable output and selection are necessary. A compromise has therefore to be struck. Divergent lines are permissible, to meet the need for enough animals, but the lines must be perpetually pruned, with the exception of a single line that, in the light of its performance, is regarded as the most favourable. All the other lines that are pruned become branch lines, with no posterity. It is probably safe to allow branch lines to run for some three generations.

The pedigree of an inbred strain will therefore look something like Fig. 4.2, where each horizontal row represents a generation, and each circle a

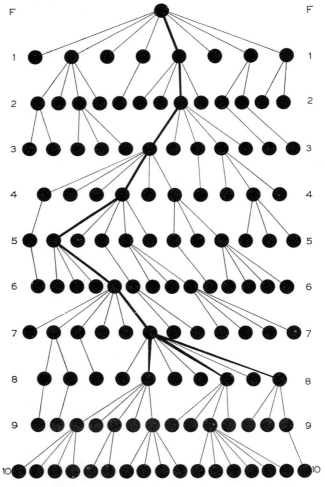

Fig. 4.2. A typical family tree for an individual inbred strain. The thick lines represent the single line of inheritance contributing to posterity, and the thin lines the short branches that do not contribute. At the eighth generation there are three possible choices; one of these will be confirmed by selection based on the record of the tenth generation, and the other two abandoned.

breeding pair. Such a pedigree only records those pairs that are used for
breeding. All the young born to any of them, but not retained for breeding
and therefore not shown in the pedigree, represent the output of the strain
available for other purposes. The field of selection at any stage in the
history of the strain is thus much larger than might be estimated by examining the pedigree chart, for the chart only shows potential breeders; that is,
the short list from which the final choice is to be made.

A colony run on these lines is still a small one, and the total output will
not be enough to satisfy the demand for animals for experimental use. Nor
is that the intention. Such a colony aims to provide breeding stock for sub-

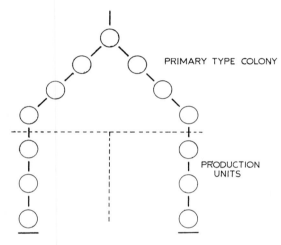

Fig. 4.3. If there are three generations at most in each branch line within the primary type
colony, and not more than three generations of sub-cultivation in the production units, then
the maximum possible number of generations separating animals in two production units is
twelve; that is, six from each limb of the family tree back to a common pair of ancestors.

sequent expansion elsewhere, about which much more will be said in Chapter 10. But if divergence of sublines is to be avoided, this subsequent
expansion must be limited to a small number of generations, enough to
ensure a sufficient number of animals ultimately produced for experimental
use, but no more. If divergence is regarded as a negligible danger within
about ten generations of separation from a common ancestor, and if for the
reasons given branch lines in the pedigree are limited to three generations,
then three generations of expansion are the most that can be allowed.
Under this system, the maximum number of generations separating an
animal in one ultimate generation from one in another ultimate generation
is twelve; in each case three generations back to the pedigree colony, three
up a branch line to the common ancestors, three down another branch
line and three more outside (Fig. 4.3). This is near enough to the limit

already stipulated. With prolific animals like mice (in which species the interest chiefly lies), even inbred mice, the ultimate numbers produced for experimental purposes will be considerable.

Suppose an inbred strain has an average litter size of n young, with a sex ratio of 1 : 1, and that on an average each female produces m litters during her breeding lifetime. A single pair could produce $\frac{n}{2} \times m$ pairs in the first generation; $\left(\frac{n}{2} \times m\right)^2$ pairs in the second generation; and $\left(\frac{n}{2} \times m\right)^3$ pairs in the third generation. This could theoretically produce $2\left(\frac{n}{2} \times m\right)^4$ young. If $n = 6$ and $m = 6$ (which are reasonable figures for many inbred strains of mice) the value of $2\left(\frac{n}{2} \times m\right)^4$ is 209,952 mice.

Of course, this figure can never be achieved because not all the animals in all the breeding generations can be put together in fertile pairs; but even half the theoretical output in four generations is a substantial number for experimental purposes, and in practice breeding beyond the third generation will seldom be necessary.

There is a corollary to the system of limited subcultivation of inbred strains that is of practical importance in the production of large numbers of animals for experimental use. In the pedigree colony brother by sister mating must be continued indefinitely, if the strain is to keep its inbred status. But in the three or four generations of expansion outside the pedigree colony, the abandonment of brother by sister mating will cause a negligible loss of homozygosity. Provided the subcultivation is not taken further than this, it is permissible to mate at random within the colony while still having in effect inbred animals. The proviso is vital, and it follows that mating, although in every other sense it may be random, must yet be between members of the same generation.

The exploitation of this for large scale production of inbred animals is described in Chapter 10.

There is always the possibility of a mutation occurring during the process of subcultivation, or even within the pedigree colony, and escaping detection. What will then happen under this system will depend on the nature of the mutation; whether it is dominant or recessive, lethal, advantageous or neutral, easily recognizable or not, and so on. Working out the consequences of the various possibilities makes an interesting exercise in theoretical genetics, but from the point of view of the uniformity of the ultimate product of breeding under the system above described, random mating during the three generations of subcultivation will be found at least as good as brother by sister, and much easier to apply.

IV. Selection

True random mating denies the exercise of selection, for selection is the choosing of matings in accordance with a programme which cannot be random (although it could be statistically randomized). In fact, some degree of selection is unavoidable in all colonies; that is, natural selection. Non-viable or infertile individuals will not reproduce, and so the field is monopolized by the fertile. If the aim is to produce a reasonable number of animals, some artificial selection will have to be applied also. The more prolific will be favoured as breeding stock for the next generation, or the more vigorous, or those displaying characteristics it is desired to enhance or fix. In other words, in any colony whose main aim is to produce animals for a variety of purposes, whether inbred or not, selection for useful characteristics is bound to be practised.

It should be remembered, however, that selection exerts its effect on characters that are heritable within a population where there is room for selection. Now, heritable characters are genetically controlled, if one accepts this as a definition of heritable. But if heritability can be allowed to include characters other than these but nevertheless transmitted from one generation to the next, then there are some heritable characters that are not genetically controlled. Such could be infections, or the sustained effect of chance accidents—for example, the provision of an inadequate diet leading to poor growth, poor development, poor lactation, with its effect on the next generation and perhaps the one after.

It follows that selection in a colony that is 100 per cent homozygous (if this is possible) will have no effect so far as genetically controlled heritable characters are concerned, for all the members of that population are genetically indistinguishable. But there may be some non-genetical characters, which offer a field of selection. The presumption of 100 per cent homozygosity in an inbred strain of long standing rests on evidence that is by no means certain, while the presence of non-genetic heritable characters is indisputable. Hence, it is reasonable to attempt selection even in inbred strains, and in practice it often helps to improve them. Since most mutations are detrimental, it will hasten their elimination and thus help to keep the colony genetically stable. If there is an irreducible minimum of heterozygosity compatible with viability or vigour, selection will also preserve this minimum and could even raise it. But this is not important provided that selection is not confined to characteristics such as vigour or productivity, but also takes into account the specific characteristics of the strain which make it useful.

For example, the cancer worker, who has a special desire for inbred strains, is interested in their incidence of spontaneous tumours, or the certainty of being able to graft tissues or tumours from one individual of the strain to another (isograft). The spontaneous tumour rate can be observed and selected for. The ability to make isografts depends on a fairly small number of histocompatibility genes. The cancer worker is not interested in the proximity of his mice to the theoretical 100 per cent, maximum homozygosity. So long as they develop tumours to schedule and the isografts take, let the genes do what they will. Selection may have produced and preserved a small residual heterozygosity, but it has ensured a suitable experimental animal.

Selection does, in fact, have a stabilizing influence, provided the right things—that is, the wanted things—are selected for. But some of the wanted things are not immediately obvious. They have to be tested for; and once again we are back to the absolute necessity of a small colony.

The most obvious criterion for selection, and one of those most easily measured is productivity, which may be defined as the number of young born and weaned per breeding female in unit time. By definition this is a function of litter size, viability of young and time between birth of successive litters. The effects of selection for maximum productivity are far-reaching.

As mammals go, mice are very prolific. They have large litters and, because of the existence of a post-partum oestrus and a short gestation period (20 days as a rule), they are theoretically capable of producing a litter about once every three weeks throughout their breeding life. To take advantage of the post-partum oestrus it is necessary that the male be kept in the cage all the time, or at least that he shall be present at the time of parturition of the female. In mice his presence is tolerated at this time, as well as during the nursing period—in fact, all the time—so that monogamous mating without separation can be safely practised. The female will then be frequently pregnant and lactating simultaneously and this puts on her a severe physiological strain.

This is the first effect, for the less robust females will not stand up to this strain, will fall down on such intensive breeding, and the offspring will thus be less likely to be selected as future breeders. To the extent that high productivity is genetically controlled, selection will favour the inherently more robust. But a relative failure under these conditions may also be due in part to inapparent infection in some individuals in the colony. If this infection is transmitted mainly vertically, that is from parent to offspring, then the infected lines will tend to be eliminated.

Tuffery (1959) has shown that, in respect of an infection that is transmitted horizontally (that is, from cage to cage) and that has an incidence of less than 100 per cent, the chance of any individual becoming infected is directly related to the time the individual spends in the colony. Hence a breeding lifetime that is shortened through intensive breeding will lessen the chance of spread of such infections.

Selection for productivity, therefore, not only favours the genetically more productive lines, but also militates against the incidence of infection, both vertical and horizontal.

There is, however, no reason to suppose that productivity is related in any way to other characteristics, which may be important in any particular strain. Hence the necessity for combining, in the selection programme, criteria of productivity with those of other wanted characteristics. But generally speaking it is hard to see how selection for productivity can ever be entirely ignored. If it is, the fertility of the strain will go down to a level that may be quite impractical and uneconomic.

V. Inheritance

A. GENOTYPE, PHENOTYPE, AND DRAMATYPE

At some time in its life the laboratory animal is destined to become an experimental animal; that is, it will be the subject of an experiment, the result of which will depend on how the animal reacts to certain procedures.

Now, an animal at all times is the result of the interaction of its genetic make-up—its genotype—with its environment. The genotype is fixed for the lifetime of the animal at the moment of conception: only somatic mutations, that is, karyotypic changes occurring in individual cells of the animal, can alter the integrity of the genotype, and the effects of these, if they occur, will be primarily localized in the cells or tissues concerned, and not generalized throughout the whole body of the animal. For the sake of the present discussion they can be ignored.

There is a sequence of interactions between the genotype and the environment. Unlike the genotype, the environment is always changing, from the prenatal environment following conception to the total lifetime postnatal experience of the animal. Environment, in this connection, is not merely the atmospheric environment, but the totality of all exogenous factors acting on the animal: food, caging, care, microflora, etc. If the total environment is considered as a single factor, then the animal under observation, which is called the phenotype, may be represented thus:

$$\text{GENOTYPE} \times \text{ENVIRONMENT} = \text{PHENOTYPE}$$

where $\times$ represents interaction (rather than summation).

If the genotype of an individual animal is constant, and the environment is changing, the phenotype must therefore also change; the animal is a different animal, phenotypically speaking, at different times. Of course it is; it grows, develops and senesces, and it may also possess good or indifferent health, and so on.

Consider a group of animals, of identical genotypes: the members of an inbred strain, for example. If they are surrounded by the same environment they must exhibit identical phenotypes. In practice, it is virtually impossible to ensure that the environment of a group of animals is identical all the time for all of them; among other factors, there will be social relationships that will create differences. In the case of inbred strains, the effects of small changes in the environment are more likely to cause significant variations in reaction to some changes among a group of animals, because the homozygous state is less well buffered, and therefore more labile, than the heterozygous. This accounts for the apparent paradox that inbred strains, despite their genotypic identity, are not generally more uniform than outbred animals, except in characteristics that are highly heritable.

Going back to the sequence of interactions between the genotype and the environment, the time approaches when the laboratory animal is to become an experimental animal. The experiment will demand certain preparations that will affect the animal through its environment, and thus to some extent transform it. Russell and Burch (1959) have expressed this in the following way:

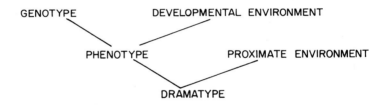

The dramatype is the experimental animal. There is no sharp line separating the developmental from the proximate environment; they are both parts of the same continuum. But the proximate environment is, in many experiments, intended to have a distinct and regular effect on the animal. For example, a rat destined to be used for assay of vitamin D will need to have its reserves of this vitamin depleted, to the point that early rachitic changes in the bones may occur. The process of depletion is an example of Russell and Burch's proximate environment; it is something specifically demanded by the experiment.

B. HERITABILITY

Some characteristics of an animal, such as coat colour, are controlled by a small number of genes, or even a single gene. The expression of this gene may be influenced greatly, a little or not at all by the environment. Other characteristics may be influenced by a number of mutually modifying genes, and the expression of these genes may likewise be more or less influenced by the environment.

Lyon (1963) gives some excellent examples of inheritance of characteristics in the mouse. In all her examples the genotype has a profound influence on the phenotype, which is little if at all modified by the environment. They are, therefore, examples of characteristics showing a high degree of heritability.

Falconer (1967) has discussed heritability in a paper that should be studied in detail; and the calculation of heritability is explained by Dinsley (1963). Their examples are, however, mostly of characteristics influenced by a number of genes; perhaps an unknown number, and unlikely to be small.

Monogenic (one gene) or oligogenic (a few genes) characteristics that are of experimental interest are likely to have a high heritability. Polygenic (many genes) characteristics are more commonly influenced considerably by the environment and are thus likely to be of lower heritability. From the genetical point of view, characteristics of high heritability are more interesting, and they find a use in certain specific fields of work.

1. Specific Genotypes

Apart from genetical research itself, the greatest use that is made of animals with specific genotypes is cancer research. In this field a large number of strains of mice, and a lesser number of strains of rats and other animals, have been developed that have special susceptibilities to tumour induction, or high or low incidences of spontaneous tumours. Most of these strains developed for cancer research are strictly inbred, and for a particular reason. In such research it is often necessary to transplant tissues, tumours or cells from one animal to another. Unless the donor and recipient animals are of identical genotype (or at least of compatible genotype, as in the case of a transplantation from inbred parent to F_1 cross) this transplant will be rejected. So it is that cancer research has been an enormous stimulus to the development of inbred strains, above all of mice.

Specific genotypes do, however, have many uses outside cancer research. There are some pharmacological reactions that depend on one or a few genes, but do not demand the high level of homozygosity of an inbred strain. A classic example of this was reported by Harris (1965). He was investigating the anaphylactoid reaction in rats injected with dextran, and found that some rats reacted, while others did not. A careful investigation demonstrated that the reactivity was dependent on a single gene in an outbred strain of rat.

Less obviously specific are the responses, chiefly in the pharmacological and immunological fields, characteristic of certain outbred stocks of animals. In some if not in most cases the investigators have become accustomed to animals of a particular stock, and possess a wealth of baseline data about them. To change to another stock would in all probability necessitate collecting a whole new set of baseline data. Now, in such instances the responses of interest to the investigator appear to be genetically controlled, at least in part, and this implies that one outbred colony is genetically different from another, and that these differences are fixed and enduring. It would also suggest that such fixed characteristics in outbred stocks are controlled by a small number of genes, for which the stocks are homozygous.

Baseline data and other characteristics include growth rates, reproductive performance, the incidence of tumours and degenerative diseases, average life span, as well as specific responses like the reaction to dextran. All these are to be found in outbred stocks, and will distinguish one such stock from another. It should be added, however, that the heritability of such characteristics may vary considerably, and their expression may therefore be more or less influenced by the environment.

2. Pathological Mutants

A special case of specific genotype is the pathological mutant. In 1956 Grüneburg published *An Annotated Catalogue of the Mutant Genes of the House Mouse*. In this he listed some 110 mutant genes that had been studied in the mouse and that were of potential value in medical research. These were only a selection from a much larger number, and they were chosen because their use was considered to be practical. Reference should also be made to Sabourdy (1970), for a much broader treatment of the same theme.

Very little use seems to have been made of the opportunities presented by such mutants, many of which resemble hereditary diseases in man. One reason is undoubtedly that medical research workers are too often unaware of the existence of such mutants, or they lack the means to produce them and have not a ready source from which to obtain them. Apart from this, however, it cannot be assumed that a mouse affected with *dystrophia muscularis* (caused by recessive genes *dydy*) will be a suitable model for studies on muscular dystrophy in man: it probably will not. Stewart (1969) showed that similar end results of genetically controlled conditions could be arrived at by different paths in genetically dissimilar animals.

Nevertheless, there is a large field of potentially useful pathological mutants, not only in the mouse but in other animals. Their production presents certainly genetical and practical problems, but these can be met in a suitable breeding unit.

C. GENETIC MONITORING

Rules for a standardized nomenclature for inbred strains of mice were first worked out and published in 1952 (Carter *et al.*) and these have been revised a number of times since (in 1960, in 1964 and in 1968; Staats,

1968). Similar principles are generally recognized in the nomenclature of inbred strains of other species. In 1970 the International Committee on Laboratory Animals (ICLA) prepared recommendations for a standardized nomenclature for non-inbred animals; and in the same year the Institute for Laboratory Animal Resources (ILAR—1970) published a report of their committee on nomenclature for outbred animals.

Despite the existence of nomenclatures that may be in some details different or even incompatible, there is a clear need for general agreement on strain and stock designations, and in time general acceptance of a single code of nomenclature will be reached. Especially in the case of inbred strains it is absolutely necessary to have general agreement, because without it there is no satisfactory way of designating a strain, or a mutant, or a gene, with unequivocal accuracy.

Inbred strains are interesting, not only because they are inbred, but also because of their genetic make-up, which confers upon them those particular heritable characteristics that make them useful experimental animals. But, as with anything that has a name intended to be descriptive, it is necessary to be sure that the name is accurate; or that the product does in fact fit the name. With laboratory animals this requires constant or periodical testing, first to demonstrate that the animals are correctly described, and second to continue to demonstrate that, through mutation, miscegenation or any other cause, they have not ceased to be correctly designated. A programme of genetic monitoring is therefore necessary in any colony, inbred or out-bred, in which the genotype is of importance. Four kinds of genetic monitoring will be considered: skin grafting, test mating, specific response testing, and biochemical methods. This is not an exhaustive list of methods, but is given by way of illustration.

1. Skin Grafting

When an organ or a tissue is removed from one animal and implanted into another, one of two things may happen. The implant may be accepted by the recipient, be incorporated into its own tissue, just as if it had always been there. Or it may be rejected, as foreign, and not compatible with the recipient's own tissues: it will not be incorporated, but after more or less time it will be sloughed off.

Not all tissues are equally sensitive, either to being transplanted or to receiving a transplant. Skin is one of the most sensitive. The acceptance or rejection of a skin transplant or graft in mice depends upon a number of histocompatibility genes located in a number of different chromosomes. If the same alleles are present in the graft as are present in the host, the graft will be accepted—will "take": otherwise it will be rejected.

Skin grafting, therefore, is a sensitive test of the identity of alleles in certain loci in a number of chromosones, all of which alleles in graft and host have to be compatible. If they are, there is a strong probability that neighbouring genes are also identical, and it is on this probability that the value of skin grafting depends.

Many generations of brother by sister mating produce an inbred strain, in which the degree of homozygosity approaches 100 per cent, and in which the genotypes of all the animals in the colony are presumed to be identical. This genotypic uniformity continues until a mutation occurs, or there is a mating outside the colony. If a single gene mutates, and it is not a histocompatibility gene, skin grafting will not be affected. But if there is a mating outside the colony, the genotype of the progeny will be grossly altered, and histocompatibility genes are certain to be involved. Skin grafts will no longer take, thus revealing a loss of histocompatibility. The same rejection of grafts will also occur if there is a mutation at a histocompatibility locus.

In colonies of inbred strains it is customary to test for continued histocompatibility, from which can be legitimately inferred continued purity of strain, by routine skin grafting. Methods of skin grafting in mice and rats have been described by Bailey and Usama (1960), by Brown (1963) and by Festing and Grist (1970). See also Parrott and Festing (1971).

2. Test Mating

In some cases the presence or absence of a desired allele is not revealed in the phenotype of the animal. A simple example of this is a sex-linked recessive gene that is carried on the X-chromosome. This will express itself in the male (XY), because the Y-chromosome has no normal allele to dominate it; but it may be carried by the female, who is normal but will transmit it to half her male progeny. The only possibility of obtaining an affected female is by mating an affected male with a carrier female. Half the male progeny and half the female progeny will be affected (Fig. 4.4).

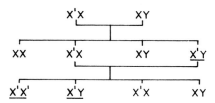

Fig. 4.4. Possibilities of the transmission of a sex-linked recessive gene and its expression. X, Y—normal alleles. X'—mutant allele; where this is expressed the genotype is underlined.

This is a very simple example of the use of test-mating procedures to work out the genotype of animals, or such parts of the genotype as are of interest to the investigator. Test mating may be used either with inbred strains or with outbred stocks. For a number of examples of this, and related genetical manipulations, the reader is referred to Lyon (1963).

All forms of test mating give information some time after the event. If the method is used in a selective programme, it will mean that a number of candidates for selection will have to be kept in the colony until the results of test mating are known. This can be wasteful.

The expansion of inbred strains, referred to earlier in this chapter, makes general use of the principle of test mating. Here ordinary (brother by sister) mating is carried out in a number of lines, all of which have a common ancestor not further back than three generations. It is on the performance of these lines that a decision will be taken which line to preserve in the colony and which to terminate.

3. Specific Response Testing

For many purposes a particular strain of animals, inbred or outbred, is chosen because it has been found to react in a particular way to one or more specific procedures. The specific reactivity of a certain stock of Wistar rats to the injection of dextran has already been referred to, and this has been shown to depend on a single allele. The only way to find out whether a particular colony possesses this allele, and is indeed homozygous for it, is to inject dextran into a sufficient sample of animals from the colony and note the results.

This is simple, monogenic, response. Probably most specific responses of interest to the investigator are polygenic and much less clear cut. Some strains of mice are useful for assaying pertussis vaccine and others are not, and this has not been shown to be due to one or even a small number of genes. But certain colonies of mice are useful for this purpose and others are not, and the only way of distinguishing between them is by testing for the specific response in question.

There are specific resistances and susceptibilities to infection possessed by different colonies of many species, and these also depend on an unknown number of genes: the situation may well be complicated by a heritability of such characteristics well below 100 per cent. The specific characteristic therefore depends, not only on the genotype, or on those parts of it that are relevant, but on the environment in which the animals are raised and kept. In the field of industrial pharmacology the suitability of the animals for the purpose is always important and sometimes critical. The pharmacologist chooses his animals on the basis of their responses to his experimental procedures, and he will be quick to detect any failure of the animals to respond in the way that animals from the same source have been accustomed to respond.

4. Biochemical Methods

Another method of checking the genotype of animals in an inbred colony has been described by Moutier (1968, 1971). In this method haemoglobin or plasma proteins are subjected to electrophoresis, and it is found that the patterns so produced correlate completely with the genotypes and can in many cases be related to specific loci. The method appears to be quite as sensitive as the skin-grafting method, and it has the advantage that a mouse can donate enough blood for electrophoresis without being damaged, as happens in skin grafting.

VI. Summary

The variety of laboratory animals available to the investigator is virtually limitless, but in fact only a very small number of species are used in appreciable numbers. Of these, the common rodents, rats, mice and guinea-pigs, are by far the most numerous; and it is in these species that, among mammals, genetical studies have been most advanced.

The consequence is that, in the common species of laboratory animals, the investigator has a choice, not just of species, but of strains within the species. These strains are of great variety, and in many cases have obvious parallels or similarities with pathological conditions encountered in man; and they are also in many cases well defined.

The use of inbred strains, especially of mice, but increasingly of other common species as well, is essential for all kinds of transplantation studies. To keep such strains, and to exploit the possibilities that they present, as well as to handle interesting mutants, requires some knowledge of genetics. This chapter makes no pretence of supplying such knowledge. But there are certain practical measures, dependent on the proper application of genetical principles, that no one who is responsible for providing well-defined laboratory animals can overlook.

CHAPTER 5

Health

I. Concept of Health and Normality

The effects of disease on laboratory animal colonies are both simple and serious. It may lead directly to deaths at any time from birth until the animal is due to become an experimental animal, so that all the effort put into its production is a complete loss; or the animal may die during the period of experimental use. In this event there will be more or less interference with the experiment, from total loss to diminution of the value of the findings. In practice, the research worker, foreseeing the possibility of intercurrent losses from disease, will probably have insured against it by using a larger number of animals than would have been necessary if he could have relied on obtaining healthy animals in the first place.

Infection may depress breeding performance, productivity or growth rate, all of which represent a reduced dividend on the investment put into the animal colony. Quite as serious, in many ways, is the fact that infection does not strike uniformly, so that there is also a reduction in the phenotypic uniformity of the colony. Thus, for experiments which require, ideally, groups of uniform animals, larger groups of infected animals will have to be used to obtain the same results, by statistical manipulation, as could be obtained by smaller groups of more uniform animals, even though none of the animals actually dies of intercurrent infection. Infection may even divert or reverse the direction of experimental results, or make extra nutritional demands, quantitatively and qualitatively. Thus, it may lead to inefficiency, frustration or misleading investigation, all of which are costly, and generally much more costly than the available remedies.

The investigator, then, needs animals that do not show any departures from health; he thinks of such animals as normal, and he expects them to remain so throughout his experiments.

There are several inherent contradictions in this concept. In germfree animals there is, by definition, no possibility of infective disease, but the animals can hardly be regarded as normal, since they lose their germfree status the moment they leave the shelter of an isolator. Indeed, in nature mammals are never to be found germfree except during intra-uterine life, and therefore the germfree state must be regarded as totally abnormal and artificial.

Much the same can be said of animals that, while not claiming to be gnotobiotic, have been raised in conditions of strict isolation, and have a limited microflora. When they are exposed to the less rigorous environment of the experimental laboratory, changes may occur in their microflora, some of which may lead to disease. The investigator does not regard the original animals as normal, possibly because they undergo changes when he starts to use them; nor will he be pleased if disease develops due to intercurrent infection in the course of his experiments. What he wants is a group of animals that are free from disease and can remain so in his experimental environment; these he will regard as the "normal, hardy animals" referred to by the Commission on Drug Safety (Report, 1964, a) and discussed more fully by Lane-Petter (1970, e).

II. Departures from Health

Disease, defined as a departure from health or from the "normality" of the previous section, may have many aetiologies: genetic, neoplastic, developmental, degenerative, nutritional, toxic or infective. Some of these may interact, and indeed no pathological condition is likely to have a pure single aetiology.

Naturally occurring genetic and neoplastic diseases or abnormalities are of chief interest to the geneticist and in cancer research. Developmental and degenerative conditions often have a strong genetic element. All these may be provoked or modified by toxic agents, which can include mutagens, carcinogens, teratogens and cytotoxins, and these are the particular province of the toxicologist. These non-infective causes of disease will not be considered further here. Nutritional diseases are referred to in Chapter 7.

There remains the last main cause of disease in laboratory animals, namely, infection. This has preoccupied the thoughts of laboratory animal breeders and users for many years, and a considerable mythology has grown up about it.

A. CAUSES OF INFECTIVE DISEASE

Among the more persistent myths about the cause of infective disease is that which may be expressed by the equation

$$H + P = D$$

where H is the host, P the pathogen and D the disease. Schneider (1970) has expressed this more realistically in the following terms:

"Medical microbiology asserts that human tuberculosis is caused by *Mycobacterium tuberculosis* var. *hominis*. No tubercle bacillus, no tuberculosis.

"Medical ecology might cast this statement somewhat differently, as follows:

"*Mycobacterium tuberculosis* var. *hominis* is a necessary condition for the appearance of tuberculosis in human subjects. It is not, however, a sufficient condition, for it has been demonstrated that human subjects may come into contact with the bacillus and *not* develop tuberculosis. Aside from the possibility of previous exposure, the list of conditions that must be appended in definition to achieve sufficiency, some of which are clearly involved but not yet clearly defined, include genetic constitution of the host and pathogen populations; dosage; past and present host nutritional history; environmental stress factors such as climate, housing, and physical work performance; and certain features of psychic health and psychic burdens."

In this definition Schneider introduces the terms "pathogen" and "stress", and lists a number of conditions that modify the simple equation.

In the attempt to provide animals free from infective disease, a disproportionate effort has been made to exclude the pathogen from their environment. In the case of gnotobiotic systems isolation is a certain protection, but to regard a whole animal house as a sort of scaled-up isolator is not justifiable. Such barrier buildings invariably belie this concept: they always admit a pathogen sooner or later, and the pathogen always establishes itself in the colony, given time. Since, according to Schneider, the presence of the pathogen is not itself a sufficient condition for disease, it would seem useful to look at the other necessary conditions.

"Stress" is a much overworked word, which has a precise meaning, but which is also used to cover all aspects of the host environment that favour the development of disease in the presence of a pathogen. Malnutrition, unfavourable aspects of the physical environment, social factors, poor handling and the burden of experimental procedures can all cause an animal that is infected but unaffected to develop lesions attributable to the pathogen. In conditions in which it is impossible to exclude with certainty, all the time, the pathogen—and this includes the vast majority of experimental situations—attention to these other necessary conditions will be rewarded by healthier animals.

Schneider also mentions dosage. There are a very few virulent microorganisms that will always produce disease in the host, even when only a single organism successfully invades. But these are exceptional; the more usual case requires a minimum or threshold number of organisms to gain entry into the host, in suitable conditions, in order to set up the disease process. The condition of the host and its environment are important; so too is the route of entry of the pathogen. But where a threshold infective dose of organisms is necessary, this may be expressed by another simple equation—this time a valid one:

$$ID = N \times T$$

where N is the number of pathogenic microorganisms, T the time during which the host is exposed to them, and ID the infective dose necessary to establish the state of infection. From this equation it is clear that by cutting down on the number of organisms, or by reducing the time of exposure, the chance of establishing infection will be lessened, even eliminated. Kraft (1967) has illustrated this principle very elegantly in relation to infantile diarrhoea in mice. If mice are kept in a cage with a filter cap, and the cap is removed for one minute every day, the exposure to a possible infective dose of virus is only 1 part in 1440 of the exposure that would be incurred with no filter cap, and therefore the chances of the animals becoming infected are reduced by a similar factor.

But if this potential pathogen does successfully invade, it may still fail to cause disease. In such cases it is often referred to as an inapparent, or latent, infection. A truly latent infection undermines the right of the microorganism to be called a pathogen; indeed, perhaps the term pathogen should never be used, or only for those very few organisms that always produce disease under all conditions. It would be more correct to say that microorganisms can, in certain circumstances, be pathogenic (with due regard to Schneider's ecological argument), but that the same organisms are in other circumstances non-pathogenic. As Sacquet (1965) has pointed out, so many of the common commensals of laboratory animals will in certain circumstances be pathogenic (but usually are not), that to exclude them all will produce an animal that is anything but normal and will not remain unchanged in a normal experimental environment. *Escherichia coli* is such an organism. It can seldom be excluded, and never for any length of time, in any system more open than an isolator; it is ordinarily a non-pathogenic commensal, but it can on occasion be pathogenic. *Pseudomonas aeruginosa* is another: so are *Clostridium welchii, Proteus vulgaris, Staphylococcus* spp. and a number of viruses. Dubos (1968) lists as "normal flora" of mice bartonella, salmonella, *Corynebacterium kutscheri*, the PVM virus, the viruses associated with epidemic diarrhoea of infant mice, and the virus of lymphocytic choriomeningitis. "All these agents can be evoked from the state of latency into pathological activity by various forms of stresses, for example administration of cortisone, excessive crowding, or malnutrition." Whether such organisms, and many others, can be tolerated in a colony, or must be certainly excluded, will almost always depend on the circumstances of the use to which they are destined to be put.

Inapparent infection is a serious problem, because it is, at least for a period, inapparent, but may yet show itself as overt disease under suitable provocation. The term "latent infection", which is often used in this connection, implies that the infection provides no evidence of its presence. But the evidence, though not obvious, may still be there, if it is only looked for. The presence of salmonellosis of low virulence in a mouse colony may give no sign of recognizable disease, but the causative organism may be cultured from the faeces or the water bottle. Or the infection may cause a slight lowering of normal healthy activity in the animals, or depress their breeding performance, to a degree that is only recognizable by one who has an expert knowledge of the normal behaviour or who measures productivity continuously. A comparison may be made with the skilled physician who can detect morbid signs at a very early stage of disease, signs that are easily missed by the less skilled.

Are such infections truly latent, truly silent, giving no evidence of their presence, or is it that the signs they give are too easily overlooked? True latency may exist, but the term is often loosely used for conditions that are not completely silent, but merely unobtrusive. The more careful the search that is made for the presence of pathogens, the fewer will succeed in concealing their presence; and the more carefully the animals are observed, the more likely are signs of infection to be revealed.

Latency is an interesting concept, but one that should be applied with great caution. It can be a cloak for incompetent husbandry or diagnostic bewilderment—who has not writhed under the impact of an autopsy report attributing death to "latent virus infection"?

The causes, then, of infective disease in laboratory animals are compound, and the prevention of disease will be helped by a combined effort not only to exclude the potential pathogen, but also to remove the other necessary conditions for its manifestation.

B. THE ACQUISITION OF INFECTION

In this context infection is used to cover all types of microorganism and parasites that may become associated with laboratory animals, whether or not they provoke or can give rise to disease. A microorganism that is present in a host animal is an associated microorganism; if it causes disease in any particular case, it becomes a pathogen. The same species of microorganism can be a pathogen in one animal and non-pathogenic in another; it can be non-pathogenic this week, pathogenic next week, and possibly non-pathogenic the week after in the same animal.

A single species of associated microorganism may be non-pathogenic by itself, but become pathogenic in the presence of another microorganism. A classic example is mouse hepatitis virus (MHV), which by itself causes no disease in mice, but in the presence simultaneously in the mouse of *Eperythrozoon coccoides* it will cause a characteristic hepatitis (Gledhill, 1962). Such interactions are by no means uncommon, and they make it difficult to classify any particular organism as a pathogen or a non-pathogen.

In any open system—that is, other than a germfree isolator—infections will be acquired, sooner or later, because peripheral barriers are not, like germfree isolators, absolute bars to microbial invasion, but only more or less effective hindrances. In such systems the inventory of microorganisms will grow, as each new species is acquired and adds itself to the list.

The process of adding to the inventory may be represented by the sloping line in Fig. 5.1. The ordinate is a measure of microorganismal burden, and the abscissa a measure of time. The microorganismal burden is the inventory of associated organisms, due regard being paid to their liability to damage the host animal by causing disease or any other departure from health. Thus, a virulently pathogenic organism like salmonella would weigh heavily on this scale, but a beneficial or neutral commensal, like lactobacillus, would count little or not at all.

At a certain level, measured on the ordinate by the horizontal dotted line, the burden becomes excessive, and the colony of animals is then too liable to disease. This point is not, however, a sharp one. The process of acquiring new microorganisms is gradual, each organism being in effect a fresh step up the line, but the line being a continuous one. Well below the dotted line the colony is of an acceptable state of health, often described as "SPF", because its incidence of disease, epidemic or sporadic, is negligible. Well above the line the animals are constantly liable to epidemics. The dotted line should, perhaps, be rather a hatched band of indeterminate width. The time spent below the line may be regarded as the colony's golden age.

The flatter the slope of the continuous line, the longer it will take to cross the dotted line: that is, the longer will the colony under consideration take to reach a point where its microorganismal burden is no longer acceptable.

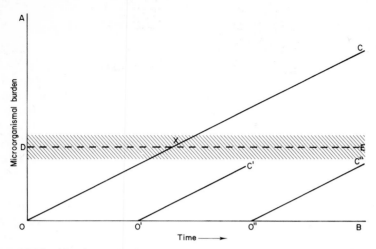

Fig. 5.1. The health of an animal colony in time. The ordinate, OA, measures the micro-organismal burden: the lower the score, the healthier the animals. The abscissa, OB, measures time. The line OC is the record of acquisition of microorganisms by the animals; where it crosses the line DE at X, or perhaps enters the hatched band, the burden becomes unacceptable. At some point before this intersection the colony should be re-founded, so that from the health point of view it can start again (O'C', O"C"). The history of the colony below the hatched band may be regarded as its golden age.

Anything that can be done to flatten the slope and thus prolong the golden age will, therefore, be a benefit to the colony. But it is essential to realize that, in any open system, the intersection of the dotted line by the sloping line is inevitable; sooner or later it must take place, no matter how sophisticated the peripheral barrier against infection (short of germfree isolators, of course).

The only way out of the dilemma is to arrange to refound colonies at some point well before the sloping line reaches the dotted line. Better hygiene, and other established measures, flatten the slope, and make it possible for a colony to have a golden age (before the intersection of the lines) of useful length. It may be that the more rigorous hygienic measures will not flatten the slope enough to prolong the golden age in proportion to the effort entailed. Refounding colonies at more frequent intervals will be found less laborious.

Today it begins to look as if the elaborate barrier systems that have been created in many laboratories are demonstrating the law of diminishing returns: they prolong the golden age of colonies, but at a price greater than that of refounding in less sophisticated circumstances. The filter rack, which is referred to on page 131, may possibly enable breeding, and more especially experimental work, to be done in conditions that are unexacting from the operator's point of view, but that provide enough of a protection of the animals to give them a golden age of useful length.

C. MANIFESTATIONS OF DISEASE

Perfect health in an animal implies that the animal carries out all its vital functions—moving, growing, reproducing, etc.—with the greatest efficiency and economy of which it is inherently capable. Disease may be regarded as any departure from this ideal, but in practice this is too comprehensive a definition, and it is more useful to think of disease as a departure from health that is associated with one or more lesions.

Disease is an abstraction, but lesions are concrete and material, and can be objectively recognized. A lesion may be destruction or damage of tissue, to the animal's disadvantage, and can then be seen by the naked eye or under the microscope. Or it may be biochemical, not discernible to the eye but detectable by chemical tests.

In this discussion it is not proposed to extend the concept of disease to abnormalities of behaviour in which there is no related physical lesion: in other words, the idea of a "psychological lesion" is ruled out.

The manifestations of disease are countless, but among the commoner signs are diarrhoea, respiratory embarrassment, poor appetite, roughening or falling of the coat, skin lesions, pallor of the mucous membranes, sluggishness and other kinds of altered behaviour, changes in the nature of urine and faeces, and so on. All these are visible signs and can usually be related to physical lesions, but minor alterations in behaviour, a slight slowing of growth or reduction in fertility, may be associated with lesions that are so slight that they may be missed. Indeed, in a colony of animals in which a high standard of health obtains, easily detectable lesions may be exceptional, yet the signs of ill-health are nevertheless present.

Tuffery (1962) has discussed the concept of non-specific ill-health; he points out that minor departures from optimal health may be recognized in the absence of easily detected lesions, and that animals showing such departures may be culled in order to preserve the high standard of general health in the colony. An animal with a dull eye, a staring coat, loose droppings or an unaccustomed pattern of movement may be in the early stages of an illness that will later become obvious. But if it is left in the colony until a sure diagnosis can be made it will be more likely, if its condition is of infective origin, to pass on the infection to other animals. The sooner it is removed from the colony, the better for the colony.

Disease of infective origin usually starts with the appearance of a few sporadic cases, perhaps confined to one or a few cages. If the infection is highly virulent, it will spread rapidly to other cages, and within days may precipitate a major epidemic in which all or nearly all the animals in the colony are affected. However, it is more likely to spread slowly at first, giving time to recognize and perhaps control or eliminate it before the epidemic stage is reached (see Fig. 5.2).

There may be several outcomes of an epidemic. Total destruction of the

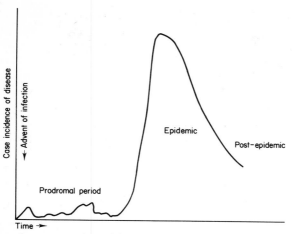

Fig. 5.2. Schematic representation of the case incidence of disease in a colony that acquires an infection, and then suffers an epidemic. There is a prodromal period, lasting days or weeks, during which cases are few and may be overlooked. This is followed by a rapid rise in case incidence, constituting a frank epidemic which, if it does not destroy the colony, gradually subsides, but never to the pre-epidemic level. The post-epidemic period is a state of unstable equilibrium, which can flare up into a new epidemic; alternatively, therapy, or the process of time, may occasionally defeat the infection, with restoration of the pre-infective state.

colony by disease is rare, if it ever happens; in the most severe epidemic there will almost always be a few survivors. A variable number of animals may die, and more become ill but make a partial or complete recovery, or remain chronically sick for a long period or for the remainder of their lives, which may be curtailed. A few, or perhaps many, will develop resistance to the disease, with or without exhibiting signs, and will carry antibodies to the infection for some time after, or for the rest of their lives. They will be resistant. In many cases, especially with bacterial infections, culling or death of sick animals will leave only resistant ones, and the infection may then disappear from the colony. More likely, the infection will persist, but the animals will be more or less resistant to it, except, perhaps, under stress when resistance may break down.

It should be clear that the actual manifestation of disease—the nature of the lesions—in individual animals is often of little or no interest except for diagnostic purposes. What matters more is whether the animals in the colony will maintain, as a colony, a good level of general health, and this can be seen better in the colony as a whole than in individual animals.

Consider 100 mice, all within 3 days of the same age and 3 g of the same weight. After one week 80 have gained 4–6 g in weight, but 20 have only gained 0–3 g. It is probable that the 20 are sick, and they should be culled. On examination, some will show no lesions, but others may give a clue to the cause of their illness. A week later, of the remaining 80 mice, 60 have gained a further 4–6 g, 15 only 0–3 g, and 5 have died. After another week, 50 have gained 2–5 g, and 10, 0–2 g, and thereafter there are no more losses. Such a health record is usually unacceptable, but it might have been improved if the 20 that did poorly the first week could have been identified by their general appearance and eliminated: the losses in the second and subsequent weeks might have been less.

Disease manifestations on this scale, affecting a considerable proportion of the colony, certainly are to be regarded as epidemic. They appear at certain times, when all the factors, infective and ecological, are in their favour, the epidemic runs its course, and the rump of the colony is left when the epidemic subsides. At that stage, the colony may be in one of the following states:

1. The infection has been wholly eliminated from the colony. This is very uncommon, but it represents, from the colony point of view, a complete cure.
2. The infection is still present, in inapparent form, but the animals have all, or nearly all, acquired a resistance to it, and can live with it. This is the common situation, and it can be followed later on by another epidemic. Meanwhile, the infection is said to be endemic, and sporadic cases of disease are to be expected.
3. Effective resistance to the infection is never achieved, and the whole colony is affected and sick; a balance is struck between host and pathogen that is permanently unfavourable to the host.

In the first situation, total elimination of the pathogen, the colony returns to its pre-infective state, in which it is uninfected but susceptible, and therefore at risk. Alternatively, the presence or elimination may have been selective, so that the animals remaining in the colony are inherently resistant: they are unaffected and resistant, and therefore at much reduced risk.

In the second situation infection is widespread, often universal, and a state of premunity may exist in which infection and the presence of active immunity co-exist in most or all animals. This state is likely to persist if the infection is universal and if it is acquired early in life, so that animals being born into the colony almost certainly become actively immune. But if the infection has an incidence far short of 100 per cent, animals born into the colony are less likely to become infected early in life, especially during their period of maternally acquired passive immunity. They may grow up with no antibodies, and thus be susceptible. When a sufficient number of susceptible animals have accumulated in the colony, the conditions exist for a new epidemic, provided the ecological requirements for an outbreak of disease exist.

In the third situation, widespread infection associated with disease, the only remedy is destruction of the colony, unless the situation can by manipulation of the environment or by other means be converted into the second situation.

D. SPREAD OF INFECTION AND DISEASE

The factors influencing the spread of disease in an animal colony are:

1. The incidence of infection in the colony: that is, the percentage of the animals that are infected.
2. The infectivity of the microorganism: that is, how early can sufficient numbers pass from one animal host to a second, and establish a state of infection in the second animal.
3. The virulence of the microorganism: that is, how well can it counter the general or specific resistance of the host and give rise to lesions.
4. The resistance of the host: this may be general, for all animals have means of defending themselves against the ordinary microorganisms of their environment; or it may be specific against that particular pathogen, either from passive immunity in the first few weeks of life, or from active immunity arising from previous exposure (including prophylactic inoculation) to the antigen.

All four factors are involved in the spread of infection. It follows that infection will spread less readily if any one or more of these factors favouring spread can be mitigated.

Reference has already been made to the formula

$$ID = N \times T$$

(when ID is the infective dose, N the number of microorganisms and T the time of exposure). ID is itself highly dependent on ecological factors. If these favour the resistance of the host, the value of ID will be raised, and so therefore will the values of N or T or both. The implication now is that animals kept in circumstances that favour resistance to infection are better able to deal with large infective assaults or prolonged exposure, than animals that are poorly cared for or subjected to environmental or experimental stresses.

It is important to understand this general principle, because it can throw light on the user's constant demand for "normal, hardy animals" (Report, 1964). To take an extreme example, in germfree animals the value of N is zero, therefore T is zero and ID must also be zero. The resistance of the animals will also be at or near zero, so that when they are removed from a germfree environment and exposed to microorganisms, infection (but not necessarily disease) must follow.

Consider now a less extreme example, in which animals are bred in conditions where the microflora is limited both in variety and to some extent in numbers too. Acquired immunity may be absent or minimal and such animals may in fact have very low gamma-globulin levels in the blood. But they do have a microflora, and their immunological machinery is not completely inactive. When they are exposed to a less protected environment than that in which they have been raised, one of two reactions is probable. The microorganisms they meet that are new, and to which they have little or no resistance, are likely to become established, either as commensals or as pathogens, and they may or may not cause disease. Or their general state of health is so good that their defence against infection with new microorganisms is successful, and new infection does not occur. In fact, both reactions are likely to occur, one with some microorganisms and one with others. But since the host's response to new invasion is a function of host resistance (one of the four factors mentioned above) a reduction in this resistance may bring later success to the would-be invader.

This is often the situation in an experimental animal house. The animals coming into it are free of an undesired infection, because they have never been exposed to it or because they have a good resistance to it. Exposure, or a lowering of resistance, will be followed by infection and, too often, by disease. In a sense some animals may be too good—that is, they not only lack a particular infection but they also, like germfree animals, lack an ability to resist it—for some experimental situations; or, to put it the other way round, some experimental situations are not good enough for, or are incompatible with, animals of a certain standard of freedom from infection.

It is important to note that, in such circumstances, it is often the "better" animal that is at a disadvantage; that is, the animal with the lower burden of parasites or potential pathogens. Animals with very low burdens have in recent years become widely available, and have been dubbed "SPF" (standing for specified-pathogen-free"), "minimal disease" and a host of other terms that seem to imply that they are superior for all purposes and in all circumstances. The multiplicity of terms in use, and the absence after so many years of an agreed definition of any of them that has a practical value, should arouse the suspicion that the whole concept is false; as indeed, it must be. What is a pathogen? How is the animal's freedom from a pathogen demonstrated, other than in a totally enclosed system such as a germfree isolator? And what pathogens is it relevant to specify in any particular set of circumstances?

This whole subject has been bedevilled with inconsequent thinking. Experience of the use of "SPF" animals has often been most unsatisfactory, and there is a call for "non-SPF", whatever that may mean; animals that are considered to be nearer the "normal, hardy animals" that are needed for so much work. For "SPF" colonies have all shown an unfortunate tendency to break down; that is, they accumulate a microfloral burden that sooner or later leads to epidemic disease.

This discussion is not intended to invalidate the demand, for specific situations, of animals that are demonstrably free of specific infection. Mice used for work on influenza must be free of Sendai virus and antibodies: mice for whole-body irradiation studies must be free of infection with *Pseudomonas aeruginosa*. Rabbits used in connection with work on typhoid and paratyphoid must have had no experience of salmonella infection. Such specific requirements are numerous and completely valid, and moreover they can be met without serious technical difficulties, although these difficulties multiply with the number of "freedoms" demanded, the number of animals needed and the size and nature of the colonies producing them. Such specific "freedoms" are those that truly merit the title "SPF"; except that the indefinable term "pathogen" ought to be dropped in favour of such terms as "Sendai virus-free", "salmonella-free", etc.

An analogy may be made with chemicals. There are analytical reagents that are necessary for many purposes, and they cost more than less pure reagents and so are not used when something cheaper will do. There are also chemicals that are guaranteed to be free of certain specific contaminants, such as lead or arsenic, but may contain others that are immaterial. Foodstuffs and drugs are strictly controlled in respect of many contaminants, many of them specific to each foodstuff or drug, but not relevant to others.

It seems more useful, therefore, to stop striving after the universal use of the elusive and indefinable "SPF" animal, and to reconsider the basic principles of health in relation to infection and how the experimental animal may best be maintained in that condition that is required by the experimenter.

Let us consider once again the simple but basic formula

$$ID = N \times T$$

The value of ID has already been discussed above. N, the number of micro-organisms to which a host or potential host is exposed, is governed by many factors.

A solitary animal has no companion from which to acquire infection, so that it is protected against all infections that are obligatorily passed direct from host to host. It is not protected particularly from infection coming in on food, bedding, the human hands that touch it, the air it breathes or the water it drinks, but many of the risks by these routes are diminished by reason of its solitary state. In many animals virus infections may disappear from solitary animals after periods of weeks, because repeated re-infection from other animals is not occurring. Thus, from the point of view of the spread of infection the solitary animal has an advantage.

Conversely, the large group of animals in circumstances of high population density is at maximum risk of infection passed directly from animal to animal, and at not much less risk of infection passed via the bedding, the dust created by the animals or via any other vehicle of infection. And it is found that, generally speaking, between these two extremes the risk of cross-infection rises both with the absolute size of the population and with the density.

There are other considerations of population, numbers and density that interfere in this relationship. For example, a solitary mouse is likely to be a stressed mouse, and monkeys and dogs often settle down better with companions than when on their own. Groups of five male rats have higher corticosterone levels in the blood than either smaller or larger groups (Gärtner, 1969). But such considerations are not important in relation to the effect of population size and density on the value of N in the equation.

These considerations favour small populations, into which, if necessary, larger populations can be broken down; and they favour low densities both of cages in the room and of animals in the cages. But considerations of expense, and of operational convenience, set a limit on the lower levels of population numbers and density, and a compromise must be accepted. Anything that can be done to make this compromise more favourable to the host will be reflected in a lower rate of spread of infection.

Among possible and practical measures are small rooms: for breeding, certainly, and for experimental use so that if possible one room does not house more than one experiment. Good ventilation will certainly reduce airborne infection, provided it is true ventilation—getting as high as possible a proportion of the available fresh air to the nostrils of the animals— and not just a circular turbulence that ensures that every animal's exhalations contribute to the inhalations of every other animal in the room.

It should of course be remembered that a group of animals in cages is, in one sense, a group of small populations, and thus different from a similar total number of animals that are freely intermingling. But these small populations are not permanent, for they are at frequent intervals cleaned out, and new groups formed or reformed, thus introducing some intermingling. There is too the mixing of the air in the room which serves all the cages; the staff handling the cages with the same hands; and the possibility of cross-infection via food or water. Thus cages in a room, and rooms in a building, are not absolute barriers to the spread of infection, but they do tend to slow it down.

Age is another factor that can influence the level of cross-infection. A new-born animal will possess some maternal antibodies—passive immunity —and will gradually acquire its own active immunity. For many infections it is at greatest risk when the passive immunity is wearing thin and the active immunity is not fully developed, either because there has not been time or because they had little or no antigenic stimulation. If stress is super-imposed at this period, the animal is placed at a further disadvantage; and this combination of unfavourable circumstances is likely to occur at about the time of weaning in rats and mice, in rabbits and in many other mammals. At this time the digestion is having to deal with solid food instead of mother's milk; weight gain is at its most rapid, social groups are often rapidly changing and, too often, the experimenter chooses this time to use his animals. If he leaves if for a week or two puberty may supervene and add a further burden.

On the other hand some diseases of a chronic or progressive nature, while present, have not had time to produce serious lesions and, histologically, the animals may be in good shape. Moreover, for every week after weaning that they are kept they cost money, in terms of space, food and care. The user is certainly in a dilemma.

Age, too, can affect the other factor in the equation, T, the time of exposure. Consider an infection of low incidence and low infectivity. The longer an uninfected animal remains in the colony the more likely is it to become infected, and in the case of many infections the time scale is measured in months. Unless the young are very susceptible (that is, the infection has a high infectivity for them) or they acquire the infection in the nest, they may well reach maturity without becoming infected. In such cases, a quick turnover of animals militates against infection. In a breeding colony this will mean early mating, early discarding of breeders, and making use of the post-partum oestrus (although there may be other equally cogent reasons against this).

On a smaller time scale, reference has already been made to Kraft's observations about the period of exposure of mice to risk of infection with infantile diarrhoea. The same principle applies to the transference of animals from one protected environment to another. Compared with the time in the breeding unit and in experimental laboratory, the time spent in transit is short, and the factor T in the equation is therefore low. If during this time N is not unduly high, ID will remain low. Indeed, it has been found that rats and mice of limited microflora, and other species, can safely be transported, by road, by air, even by rail, from one place to another, in open wire cages, with negligible risk of picking up any serious infection on the way.

III. Prevention and Elimination of Infection

To obtain a really healthy colony of animals it is possible to adopt two different courses. One can either start with clean breeding stock and maintain it in a clean environment, or one can try and rid a colony that is not known to be clear of its infections. In practice, a measure of both is necessary.

Either method depends on providing a clean environment, if continuing health is to be maintained. Here the importance of hygiene is paramount. The peripheral barrier against extraneous infection must be as impenetrable as ingenuity, discipline and circumstances will allow. And because barriers are difficult to keep inviolate, internal barriers against cross-infection within the colony are also desirable. Thus, a large colony may be subdivided into mutually isolated sections, limiting the possibility of spread from one to another.

It is, however, of little value to erect a sound peripheral barrier if infection is already present inside it. To deal with this situation it is necessary to manage the colony in such a way that foci of infection—a sick animal and its contacts, for example—are removed at the earliest possible moment. Here culling is all-important, together with more hygiene. One measure that is singularly ineffective in dealing with an epidemic is wholesale slaughter without a reorganization of the system of management. If this is not done, in the light of the way the epidemic developed, the whole sad story will be repeated.

A. HYGIENE

The aim of hygiene in the animal house is to reduce the value of N in the equation (p. 66)

$$ID = N \times T$$

by removing materials that may harbour large numbers of potentially pathogenic organisms. It is also to make the animal room, and the cages, inoffensive both to the animals and to the staff, and to preserve the fabric of the building and equipment.

All this is summed up in one word: cleanliness.

The importance of hygiene has already been anticipated in the foregoing remarks. The possible routes of the spread of infection are many, and to block some of them while leaving others wide open is unlikely to have much influence on the health of the colony. For example, it is pointless to go to great lengths in autoclaving cages if the practice is to scrape out the cage trays in the animal room, thereby producing a cloud of infective dust in the immediate vicinity of the animals. Sparkling cleanliness in the animal room avails little if visitors from outside, who may have had contact with infection elsewhere, are freely allowed to enter and carry infection with them.

A simultaneous attack on all possible routes is essential, even if complete certainty of exclusion cannot be achieved in any of them, because one is dealing with a dynamic situation. On the one hand there is the virulence of the pathogen and the massiveness of its assault, both of which contribute to the spread of infection. On the other hand are the obstacles it has to overcome to gain a foothold in the colony, the last of which is the resistance of individual animals.

The suspected presence of a virulent infection in the vicinity of an animal house is a matter of the utmost gravity. Penetration of the peripheral barrier against infection is much more likely to happen if virulent infection is present in a neighbouring animal house or wild rodent population. Physical isolation, vermin-proof buildings and vermin-free surroundings are therefore all highly desirable, if not absolutely essential. A massive invasion by a pathogen of only moderate virulence may also succeed in producing disease in circumstances where a less massive assault would fail.

Air can carry infections, especially if it is likely to develop a high bacterial content, such as recirculated air in the ventilation system unless it is well filtered. But for most purposes air is an excellent diluent for airborne infection, and the installation of dust filters will go a long way towards cutting out danger from this source.

Pests are potent vectors of infection. There is no excuse today, with the vast armoury of insecticides available, for tolerating the presence in an animal house of a single fly, bug, cockroach, flea or louse. Mites are more difficult to deal with, but miticidal agents can be applied and, properly used, are effective. The aim with arthropod pests should be complete freedom from infestation. This is attainable, and nothing less should be regarded as satisfactory.

Rodent pests are also not to be tolerated at all. The structure of the animal house and the system of receiving stores, especially foodstuffs, should rule out any possibility of entry of wild rats or mice.

Food, water and bedding all come into intimate contact with the animals, and all are potential vectors of infections. The possibilities presented by food and bedding for conveying infections, microorganismal, helminth and arthropod, are considerable, and sterilization or other appropriate treatment of these vectors is essential. Autoclaving of food and bedding is practical and certain, but consideration may also be given to pasteurization of food, which will kill all vegetative pathogens (and virtually none are spore bearers), and fumigation of bedding. Water from the mains, fit for human consumption, may be acceptable, but in certain circumstances it may require special treatment by acidification, chlorination, ultraviolet irradiation or other means. Gamma-irradiation is another very effective but more expensive method of sterilizing food and bedding; it is considered in Chapter 7.

Cages and utensils, such as water bottles and tubes, food hoppers and the like, are easily sterilized, by autoclaving, treatment with live steam, boiling or chemical disinfection. Cleaning in a mechanical washer using detergents and a hot-water rinse is for all practical purposes the equivalent of sterilization, provided the period of exposure to these conditions lasts for several minutes. Whatever method is used should combine disinfection with simple cleaning; that is, the physical removal of dirt, which provides harbourage for proliferating bacteria and concealment for arthropod pests and their eggs.

With all other channels of infection blocked, there still remains the staff working in the animal house. They cannot be sterilized, but their capacity for being vectors of infection can be greatly reduced by offering them shower baths and a complete change of clothing on entering the animal house, and every incentive or compulsion to use these facilities. But if a regular routine of taking a shower and putting on fresh clothes on every occasion of entering the animal house is not feasible, then something less is still worth aiming at. Washing of hands by the animal-house staff, and by those who have reason to visit the animal house, particular attention being paid to scrubbing the finger nails, which should be kept short; frequent changes of clothing, including the provision of suitable clean working overalls; the changing of footwear and the provision of disinfectant trays at the entrance to every animal room; and the regular screening of all personnel working in the animal house to exclude carriers of pathogens; all present opportunities for preventing the introduction of infection by human vectors. Visitors who have no business in the animal house should be made explicitly unwelcome. If they must come in, their personal decontamination must be at least as strict as that of those regularly working there, and if there is any suspicion of their having had recent contact with infection elsewhere, the health of the colony may depend on their not entering under any pretext.

Lastly, foreign animals, that is, animals introduced from outside, should never be brought in except by hysterectomy or when they have been proved beyond doubt to be free of unwanted infection.

All these precautions should be regarded as minimal and to be continuously applied. If they are combined with good husbandry and gleaming cleanliness throughout the animal house, the balance will be tipped very markedly in favour of the animal. The test of good husbandry is that it promotes good health and productivity, and contributes towards uniformity of environment—a uniformly good environment. Promotion of good health implies the reduction of infection to vanishing point, not only by excluding pathogens from the colony but also by ensuring that the colony as a whole, and even the individual animals in it, offer the poorest possible welcome to any pathogens that should get a foot inside the door.

B. CULLING

A cull is defined in the *Concise Oxford Dictionary* as an animal removed from the flock as inferior. No serious domestic animal breeder will keep an animal that is, in his view, inferior; partly because it is uneconomic to continue to feed and house it, and partly because it may be a source of danger to its mates. The good husbandman is ruthless, for he knows, often from bitter experience, the damage that inferior animals can do to his flock. Therapy may be possible in some cases, but for laboratory animals it is the exception. Elimination, which for laboratory animals means slaughter, is the only proper course in most circumstances.

But what is an inferior animal? Obviously, one that is dying or overtly sick, perhaps unthrifty or lacking those characteristics that are considered useful or desirable. But between the moment of infection and the manifestation of disease or death there is an interval of time, measured occasionally in hours, more often in days, and sometimes longer. At some point in the course of this interval, the infection reveals its presence, sometimes clamantly, sometimes coyly. The speed with which the infected animal is culled depends on the early recognition of the fact that it is infected, and the earlier it is culled the less chance has it of infecting its mates.

Good animal breeders and husbandmen have known all this from time immemorial, long before anything was known about the causes of disease, and they have not waited for obvious manifestation of disease to cull. The animal that dies on Friday was ill on Thursday and should have been culled on Wednesday—with due allowance for the temporal progress of infection —and the practice of systematic culling in fact depends upon being wise on Wednesday. This calls for a skill that can only be acquired by those in direct and constant touch with their animals, for the signs that doom an animal as a cull may be no more than a slight change in behaviour from the previous day. But it is a real skill, one that can be shown to work in any colony where it is applied, and the application of a systematic programme of culling can be counted upon to reduce deaths from endemic disease and residual ill-health.

A reliable means of diagnosis is more than half the battle against a pathogenic infection, for to diagnose is to be able to eliminate. Culling, and the criteria—often indefinable—on which it is based, are in fact a diagnostic test; a diagnostic test of that elusive entity, ill-health.

In discussing the health of farm animals, Beveridge (1960) has made much the same point. He has pointed out that: "In the past we have talked of health in terms of mortality and morbidity, the latter being based on clinical examination. This is no longer sufficient. We have to think of health in relation to production and measure the effects of disease on food conversion and growth rate, milk or egg production, which are much more sensitive indicators than clinical examination, or even laboratory tests."

Now, it is unusual to measure milk yields or meat production in laboratory animals, but there is no difficulty in measuring food conversion ratios —that is, the live weight gain for every gramme of dry food eaten—and this is a valuable indicator of health. An indirect measure of both milk yield and food conversion efficiency is the growth rate of young in the nest up to weaning. A poor performance here often spells incipient trouble, whether the animals be mice, kittens or any other laboratory mammal.

When an organism that is capable of causing an epidemic gains access to an animal colony, the course it takes may be represented by the line in

Fig. 5.2 (p. 71). There is a period when only a few animals or cages are infected, and if disease develops at this stage it is often atypical, or it is easily missed because it is not expected.

The case rate of clinical disease, therefore, is very near the baseline, and will remain there for some days or weeks, according to the nature of the infection and the colony's response to it. The next thing that happens is a small outbreak of disease, perhaps a few cages together, or sometimes cases scattered over a large part of the colony. This may die down, or it may go on in some instances to a sharp rise, and so to an epidemic. The subsequent history of the colony is likely to take one of the courses described on p. 72.

It is during the prodromal stage, when clinical cases are rare and atypical, and before the epidemic stage has been reached, that culling can be most effective, but since the infection at this time is seldom very apparent, culling may be unpractical. It would automatically take place if, in a large colony, there was a programme of regular total turnover of breeding stock, and this in fact is an efficient way of maintaining breeding units in a good state of health. This aspect is discussed in more detail in Chapter 10.

An excellent discussion of culling will be found in Tuffery's paper already referred to.

C. THERAPY

The greatest danger of spreading infection in an animal colony is an infected animal in that colony. Whenever possible such an animal should be identified and removed, as has been explained in the previous section. But there are occasions when identification of the infected animal is not feasible; or when the number of possibly infected animals is so large that to remove them would destroy the colony; or when the infected animals, which may be the subject of experiment, are so valuable that this destruction cannot be allowed.

These cases must be regarded as exceptional, but unfortunately they are not rare exceptions. While therapy—that is, treatment of individual animals to cure their infection—is normally a policy not to be contemplated in an animal house, in these exceptional circumstances it is unavoidable. It must be realized, however, that therapy may so affect the animals that, at least for some purposes, they will be rendered unfit as experimental animals.

Infections may be attacked by means of bactericidal drugs, antibiotics or by immunological means. Certain protozoa—coccidiosis in rabbits and poultry, for example—and helminths may be controlled by chemotherapeutic means; often by adding a coccidiostat or an anthelmintic to the drinking water. While such treatment will frequently reduce or eliminate the disease caused by these parasites, it seldom completely eliminates the infection, and cessation of treatment is likely to be followed by recrudescence of disease.

Much the same may be true of the use of antibiotics to control infection, which in this case is usually bacterial. Broad-spectrum antibiotics, such as oxytetracycline, are most commonly used, and they will kill a wide variety of microorganisms in the animals receiving the treatment. They may indeed sterilize the gut contents, or leave it with only a few antibiotic-resistant organisms, which are then likely to overgrow and kill the animals from another cause: the cure being in such cases worse than the disease. Antibiotics, if persistent, may also render the animals so treated unfit for their use in experiment. It is essential that an experimenter who is offered animals that have received any form of chemotherapy or antibiotics should be informed, so that he can judge for himself whether the animals are suitable for his use.

In some species the incidence of certain virus infections is such a likely and serious hazard that protection must be given as a routine. Distemper complex and leptospirosis in dogs, enteritis and panleucopenia in cats, are examples of such infections, and it is customary, and even necessary, to give all the dogs and cats in a colony protection. This is usually done by vaccinating them at appropriate times of their lives, in order to produce an active and persistent immunity to these devastating infections.

Apart from therapy of the kind indicated above, treatment of laboratory animals for disease is generally to be deplored, and in any event it is seldom worth doing. The sick animal is a danger to its mates, and should be removed and killed.

D. HEALTH MONITORING

It has to be accepted that the state of health of any animal colony is at permanent risk of undesirable infection. Since infection is not synonymous with disease, it must also be accepted that initially all infections may at some time and to some extent be inapparent, or at least not be made apparent by gross and obvious disease. Indeed, since some infections only produce obvious disease when the animal is stressed in some way, apparently healthy but infected animals may leave the breeding unit or stock room and break down when subjected to experiment.

It is necessary, therefore, to have a programme of quality control in all animal units. Such a programme can be considered under four main headings, namely microbiological, pathological, demographical and user feedback.

1. Microbiological

Now, an obvious method of finding out what infections are residing in a colony is to take a representative sample of animals from that colony and ask a microbiologist to list all the microorganisms they contain. There are, however, two difficulties here. First, what is a representative sample?

Consider an infection that is present in 10 per cent of the animals in a large colony. To have a 20 to 1 chance of picking it up from a sample, it would be necessary to examine 200 animals. Consider besides that the laboratory cannot expect to obtain a positive culture in more than 25 per cent of infected animals examined—a not unreasonable assumption in regard to some infections—and it will be necessary to examine 80 animals in order to have a 20 to 1 chance of obtaining a positive result. Consider also the microbiologist whose services are being utilized: is it really very interesting for him to process a large number of animals in order to make an inventory of their microflora? In practice, the work will be passed along to a junior technician, and the expectation of false negative or false positive results will be high.

A second difficulty is knowing how often such a screening should be carried out? Animals picked out on 1 January will perhaps be reported two weeks later: but even if the sample size is adequate, and the micro-biological technician perfect, the report will only indicate what the state of the colony was on New Year's Day. Perhaps monthly checks will give a reasonably good current report on the microbiological state of the colony, but when the number of animals and the volume of laboratory work are added up, the cost becomes formidable. Indeed, the gap between what is scientifically necessary to obtain useful results, and what is economically sensible, is for many types of infection too wide to be bridged. Another approach is necessary.

The nature of the gut flora in laboratory animals is often of considerable interest and importance. This flora is not usually defined in its entirety, because there are many organisms that may be pure commensals, are present in insignificant numbers and appear never to have any effect on the animals: they can be ignored. But a small list of common, important and potentially pathogenic organisms is not hard to compile, and this is an interesting exercise for the microbiologist. Periodical checks of this limited definition of gut flora, using small samples of animals, is neither expensive nor tedious, and it is useful.

A special case exists when interest is centred on one particular organism—for example, salmonella—the presence of which may be responsible for pathological signs in the colony. A search for a particular organism is both more interesting and more profitable than a general screening, and less costly.

It has so far been assumed that the microbiological methods entail culture of the organisms being sought. But many microorganisms may be more readily detailed by serological means, because they provoke the production of specific antibodies in the host. Viruses are a particularly good example of this: and because so many viruses are, at least part of the time, present but inapparent, their incidence within the colony is likely to be at or near 100 per cent. Refined methods of serological testing for the presence of specific antibodies will therefore reveal the virological status of the colony with a high degree of confidence. Moreover, the methods used, in skilled hands, are neither so subject to error nor so expensive as bacterial culture methods.

One particular case for straightforward bacteriological methods is in monitoring methods of sterilization or pasteurization. No special techniques are needed here: the autoclave can be tested by using indicator organisms, and its function routinely checked. Routine checks can also be carried out on the food, the bedding, the air (by means of slit samplers, or other methods) and the hands (and other parts of the body) of the staff. The regular monitoring of germfree systems, in isolators, is another case for microbiological screening, but it will often be confined to searching for one or a small number of specific organisms. In a colony with an almost perfect health record a routine search for specific organisms may give the first evidence of the advent of infection (see Fig. 5.2, p. 71), especially if this is not soon followed by an epidemic outbreak of disease. In summary, therefore, microbiological examinations are useful, and economically practicable, in the occasional definition of gut flora; in *ad hoc* searches for specific organisms or for the cause of specific lesions; in monitoring methods of sterilization, pasteurization, filtration, etc; and in specific serological screening. In contrast, the routine microbiological screening (especially by cultural methods) of samples of animals is likely to be unreasonably costly, and in many cases statistically meaningless and of very limited value.

2. *Pathological*

In any animal colony spontaneous deaths occur, and there must always be a cause for them. This is not to say that the cause is always known; indeed, too often the cause is obscure, and the report on the animal may be that it died of ill-health, unspecified. This in itself is useful, since a rise in the number of deaths of undiagnosed cause may give advance warning of an impending epidemic. To report, fairly and honestly, that an animal submitted for autopsy had no lesions that would account for its death is more valuable than a wild guess at a possible cause—such as a hypothetical virus or a problematical stress factor.

In a well-conducted animal colony the rule is to examine every death and every cull at autopsy. No immediate benefit, in return for all the work entailed, may be gained, but a subsequent examination of the records may be revealing. Simple phenomena like Monday morning mortality, due to irregular staff arrangements over the weekend, are easily spotted and soon corrected. In the course of time an examination of autopsy reports will also reveal other correlations: with a change of diet or even a new batch of the old diet; with the weather or a breakdown in the heating or ventilating plant; with a modification of the system of management; and perhaps with something totally unexpected. It should never be forgotten that some infections may be first recognized by the shape of their mortality or morbidity curves, just as some fevers have characteristic temperature charts, and thus one is led to a correlation between a given infection and an otherwise inexplicable autopsy finding.

All spontaneous deaths should be autopsied. This rule applies, even if the number of deaths is considerable, for it should be the duty of someone— a suitably qualified technician—to inspect every animal that dies. In the case of putrefying carcases, or those that have been partially eaten by their cage mates, mere superficial inspection may be all that is necessary or possible: but the fact that their deaths have been examined means that they can be recorded, as of course they must be.

Moribund or recently dead animals can be opened up in the laboratory or on the autopsy bench, and examined at least macroscopically. This is often all that is necessary: a midline incision from jaw to pubis, a quick look at lungs, heart, liver, spleen, kidneys, stomach, intestine, reproductive tract, serous cavities, upper respiratory tract, coat, eyes, teeth, limbs and tail; a practised technician can carry out many such autopsies in a very short time, and in most cases there will be one or two obvious lesions to record, or nothing at all. A single-line entry in the autopsy book is the end of the story.

Sometimes, however, lesions are seen or suspected that need further investigation. Small necrotic patches on the liver of the mouse may suggest Tyzzer's disease, but the diagnosis can only be confirmed—in isolated cases, at all events—by histological examination of the liver, in which can be seen the characteristic focal necrotic areas with pleomorphic red staining bacteria round the edge.

Histological examination of tissues or organs removed at autopsy is often of the greatest value, and the preparation and examination of a slide are usually much less expensive than putting up a culture. It also has the advantage that a slide, once made, is a permanent record that can be re-examined in the future and even sent to a colleague for a further opinion. It can be concrete evidence of a pathological condition of any kind, whether of infective origin or not, and it may throw light on deaths that, macroscopically, are puzzling.

In summary, then, all spontaneous deaths must be examined, however superficially, and recorded. If lesions are seen at autopsy they must be identified and, when necessary, submitted to histological examination. All records must be kept, because valuable information may come out of them at some time in the future.

3. Demographical

Demography, a term which has not been much used in relation to laboratory animals, means "vital statistics, illustrating condition of communities" (*Concise Oxford Dictionary*), and one of its special references is to public health.

In any animal colony records have to be kept. In a breeding unit these will include records of matings, births, weaning rates, deaths, culls (both of unfit animals and of unproductive breeders), productivity, weight gains, incidence of abnormalities (tumours, mutations, degenerative conditions, etc.); and these records can be plotted on graphs to show trends in the colony. For example, an insidious falling off of fertility will show up in the slow decline of young produced per female per week, and it may be the first sign of a vitiman E deficiency. Failure to make the expected weights at given ages, or a wider than usual scatter of weights among animals of the same age, is an early sign of an infection that has not yet become clinically evident. As Tuffery (1962) has shown, cull rates and death rates in a colony tend to be inversely related. If a steady and negligible death rate can be maintained only by an increased cull rate, it suggests a subclinical condition of some kind, probably infective.

In a group of animals under experiment, abnormal or irregular reactions to treatment may be the clue to subclinical infection, the prelude, perhaps, to an outbreak of disease. In long-term experiments the value of the individual animals rises with the time they have been under experiment, and a culling programme that would be appropriate in a breeding colony may be impossible in an experimental group. But the clues indicating that all is not well with the health of the group may make it possible to eliminate a small number of the animals, for the sake of preserving the rest, and it may also be possible to make allowances for the unwanted variable, disease, that has crept in.

Generally speaking, not enough use is made, especially in breeding colonies, of demographic observations. Technicians do not like to see their animals die, and it is so much easier to put the carcase, or what remains of it, in the waste bin, and not bother to record it. A bad week's production may seem like a reflection on the technician, who convinces himself that next week will be better. But many if not most animal technicians are very conscientious about sending their dead, moribund or culls, out for autopsy, and yet they will not be encouraged to continue doing so if they do not get reports back about what is found.

Assuming that the animal technicians send out all the animals they should, and the technician carrying out the autopsies prepares prompt and accurate reports and passes them back to the animal house, their combined efforts may still be to no purpose if the week-by-week trends are not studied.

In summary, then, records—the right records—are vitally important, especially in a breeding colony, if disease problems are to be anticipated. And it may be added that such records need not impose a serious burden on the staff; not by any means, if the most meaningful observations are asked for, and their physical recording made as simple as possible. But a word of caution: the making and recording of demographic observations should be regularly revised, not less often than twice a year, because there is a tendency to accumulate new observations that supersede the old—but the old will still be collected unless they are positively cancelled or countermanded.

4. Feedback

Feedback in biological systems is defined as "the carrying back of some of the effects of some process to its source or to a preceding stage so as to strengthen or modify it" (*Concise Oxford Dictionary*).

In a breeding unit inapparent infection may be present which will not become evident as a result of any of the stresses of reproduction and growth. Yet such an infection may be very serious under certain types of experimental stress. A classic example is infection of mice with *Pseudomonas aeruginosa*, which in most circumstances does not embarrass them. But if the mice are irradiated they will succumb to a *Pseudomonas* septicaemia. Rats may carry *Mycoplasma pulmonis* in the middle or inner ear, or in the nasopharynx, with little or no inconvenience even in conditions of intensive breeding. But if these rats are used for certain types of inhalation experiment, especially when damage to the pulmonary epithelium or lung parenchyma may be expected, then this mycoplasma will invade the lungs and cause severe, even fatal bronchopneumonia or bronchiectasis.

Less dramatic are those cases of a regular supply of animals to a laboratory, which have over a period of time given uniform and repeatable responses to certain experimental procedures, and which unexpectedly give aberrant results. This is not always due to incipient infection. In one instance male mice had been supplied every week to a laboratory where they were used for assessing a number of tranquillizing compounds. One week the responses of the mice were reversed, to the consternation of the laboratory. No infective cause was found, but it was discovered that, as a result of an act of carelessness on the part of the technician breeding the mice, a few females had accidentally been mixed with the males. The effect on the males was striking.

It should be clear that the breeder who does not seek to obtain information from the user of his animals is forfeiting a valuable aid to quality control: in some respects the most valuable, because the quality of his animals is ultimately judged by their suitability for the user.

This point has too often been overlooked. The user, as has already been pointed out, wants an animal that will respond to his treatment, and to nothing else. In this he is seeking an unattainable ideal, but it should always be the object of the breeder to aim for that ideal. A tough guinea-pig that has been bred in good backyard conditions is likely to have been exposed to a number of infections, any of which might have caused overt disease; but it can develop a resistance to them all, if they are not overwhelming, so that when it becomes an experimental animal it is not likely to succumb to the first dose of intercurrent infection that it meets. The same guinea-pig may be totally unfitted for work in which previous exposure is a positive disadvantage.

What sort of animal is the breeder to produce, or the experimenter to use? Clearly, it depends on the use, and the breeder must be guided by this, rather than by an arbitrary set of characteristics, some of which are quite irrelevant to use but will add to the cost of breeding.

This is not an argument in favour of animals of poor backyard quality, carrying an unknown burden of infection which may explode at any time into epidemic disease. It is rather an argument in favour of realism. The animals are destined for experimental use: their quality must therefore be judged in this context. It is for this reason that, in quality control, feedback is of importance. Perhaps it is the most important of all the quality-control measures, because the others subserve it.

E. DERIVATION

The ordinary laboratory animal is an association of the host animal— mouse, rat, guinea-pig or whatever it is—with a number of lower forms of life, from metazoan parasites and commensals, through protozoa and bacteria to viruses. It is these associated organisms that may, and often do, give rise to disease, or interfere with experimental procedures; and yet many of them are almost ubiquitous and are difficult to separate from the host animal.

Ecto- and endo-parasites are generally susceptible to parasiticidal treat- ment, but not always so. Mites in mice (*Myobia* and *Myocoptes* spp.) and pinworms in mice and rats (*Syphacia* and *Aspicularis* spp.) are certainly susceptible to parasiticidal treatment, but a total kill is difficult or impossible to achieve by these means, and re-establishment of the infestation occurs once treatment is stopped. Protozoal infections, such as coccidiosis, behave in much the same way.

Some bacterial infections respond dramatically to antibiotic treatment, and Van der Waaij and Sturm (1968) and Van der Waaij (1968) have reported sterilization of the gut of mice by this means. There is no evidence that immunotherapy can eliminate virus infections, although it may be a useful prophylactic. But since ordinary animals are associated with such a variety of parasites and commensals, elimination by all these means— dippings, drugs, antibiotics, vaccines, etc.—simultaneously is not really practical. There is, however, a better and simpler method.

The foetus *in utero* is, in effect, in a germfree isolator, the placenta and membranes being a very efficient barrier against infection carried by the dam. It is true that certain infections can penetrate this barrier, such as *Toxocara* in dogs, and a few viruses, such as lymphocytic choriomeningitis in mice, but not many more. Techniques evolved in germfree laboratories have made it possible to raise animals delivered just before term in aseptic conditions, and hand-raised in an isolator away from the dam. Such animals will only be infected with organisms that can pass the placenta, and for most purposes these are negligible and can be ignored. In this way the germfree animal is established; that is, the animal that has been separated from its associated fauna and flora.

Once a germfree colony has been established by the tedious process of hysterectomy and hand-raising in an isolator, it can be developed by normal breeding within the isolator, and the germfree state will continue until there is an accidental contamination through breakdown in technique. Moreover, such a colony can be used for the purpose of establishing, without the problem of hand-raising, new colonies of other strains of the same species, or in a few instances of other related species. This is done by carrying out a hysterectomy on a dam of the strain it is desired to render germfree, passing the uterus with its contained foetuses into the isolator, with full aseptic and antiseptic precautions, and fostering the pups on to a foster dam within the isolator, who has had a litter during the previous two or three days, so that she can bring up the fosterlings in place of her own.

So long as the germfree colony remains germfree this process can be repeated indefinitely. One germfree colony can then be the means of founding a multiplicity of daughter colonies, of the same, or of different, strains. The combination of hysterectomy and fostering properly carried out can in one operation separate the host animal from its associated organisms. But there are certain practical steps to be taken if this operation is to be successful.

1. Hysterectomy

Hand-rearing will not be considered here. The technique has been amply described elsewhere (Davey, 1959), and for purposes other than germfree research it is seldom if ever necessary, because germfree animals are always readily available from laboratories regularly raising them.

For a hysterectomy derivation, it is necessary to have a donor dam, who may be carrying a number of unwanted organisms, and a recipient or foster dam, who lacks unwanted organisms but may be the wrong strain or genetically unacceptable. The purpose is to transfer the progeny of the donor dam, while they are still contained in the uterus and thus not infected with their dam's microflora, to the foster dam, whose microflora either does not exist (if she is germfree) or is acceptable. The progeny from the donor dam will thus be raised as germfree or with an acceptable microflora: they will, to use a common phrase, be "cleaned up".

Now, the donor dam carries an unknown assortment of infections on her body surface, in her alimentary and respiratory tracts, and perhaps in various tissues including the blood. Even the cavity of the uterus may be infected, for example with mycoplasma (Tregier and Homburger, 1961), although such infection may not penetrate the placenta and foetal membranes. It is thus not quite accurate to regard the uterus as a natural germ-free isolator: rather it should be the placenta and membranes within the uterus.

To perform a hysterectomy, a dam within a few hours of the natural termination of pregnancy is chosen. It is possible to carry out observations of mating, so that the duration of pregnancy is known in days and hours, and this is common practice. But the duration of pregnancy, even within the same strain, is variable; a mouse may go for 17–21 days, and a rat a day or so longer, and this is largely unpredictable. Of mice from the same colony mated on Day 0, at Day 19 some may have had their litters naturally, some may be suitable for hysterectomy, and some may not yet be ready. This method is not, therefore, entirely satisfactory, and it is also rather troublesome.

At least as good a method, with a success rate in skilled hands just as high as that from observed matings, is to palpate heavily pregnant females for signs of imminent parturition. Different species exhibit different signs; guinea-pigs show a marked separation of the pubic symphysis a few days before term, which is easily felt. Rats and mice are not difficult. The foetal movements can be felt in both species, and in rats at full term the foetal limbs can also be distinguished. The foetal limbs of mice are palpable, but with less facility, while the heads of foetal mice and rats are readily palpable at term.

Perhaps one of the most useful signs in both mice and rats is the feeling of complete relaxation of the abdominal muscles of the dam when she is within 24 hours of term. She allows detailed palpation of the uterus and its contents with no resistance. Indeed, a positive and invariable sign that she is more than 24 hours off full term is a tightening of the abdominal muscles when she is palpated and, almost diagnostic, an attempt, often a savage one, to bite. This sign is particularly valuable in the case of a dam with a very large number of foetuses, giving her great abdominal distension which might lead one to think she has almost reached the end of pregnancy. If she bites, she has not.

A suitable donor dam having been chosen, she is painlessly killed by cervical fracture. Place her on a rough surface and, in the case of small rodents, lay a rod across the back of her neck, pinning her lightly to the surface. Before she has time to struggle, raise her posterior end and hyper-extend the neck, pulling at the same time to separate the cervical vertebrae after fracture. Speed is essential.

From now on the viability of the foetuses varies greatly with the species. In the case of rats and mice, they will live as long as 15 minutes between the death of the dam and delivery and resuscitation.

Various procedures have been recommended for the operation of hysterectomy. The dam may be shaved over the abdomen and painted with iodine before killing; she is then laid out on the operating board, covered with sterile adhesive drapes, incised by thermocautery, and the uterus delivered through the incision by aseptic technique. Or the unshaven dam, after killing, may be immersed in strong disinfectant, such as 1 or 2 per cent iodophore solution, laid out on the operating board, being all the while well drenched with disinfectant, incised by scalpel or scissors, and the uterus delivered into a bath of disinfectant.

Whatever technique is employed, the uterus should be separated from its adnexae and the lower end, and perhaps also the tubes, tied or clipped off, before it is placed in a container of disinfectant.

The disinfectant in the container should be at $39°C \pm 1°$: this is important for the survival of the foetuses. It is transferred to the isolator or room in which is the foster dam, by passing it through an entry tank of disinfectant. Once inside, it may be opened up; but it is important to remember that although the outside of the uterus and any blood that may be escaping is sterilized by the disinfectant, and the foetuses within their membranes are also germfree (except for the very few viruses that may pass the placenta), the muscle and the lining of the uterus, being maternal tissue, may be infected. It is therefore a wise precaution to open the uterus and deliver the foetuses below the surface of the disinfectant, so that no maternal tissue is exposed above the surface, and the foetuses when delivered are well washed. This "wet hysterectomy" technique has been described elsewhere (Lane-Petter, 1971) and is necessary particularly to prevent passing on mycoplasma infection in rats, in which this infection is commonly found in the uterus.

While still immersed in the disinfectant the foetuses are separated from their placentas, and then removed to be dried off, resuscitated, and given to their foster mother.

If the pups on delivery are rose pink rather than pearly grey, they are premature and will not survive.

Resuscitation is carried out on a warm surface: a heated table with several layers of paper towel or tissue on it, or more simply a flat rubber hot-water bottle or bed warmer, also covered with paper tissue. At first the pups will be blue or grey and inactive, but shortly after coming out into the air and being dried off they will begin to breathe. Drying them with a cotton swab is often enough to stimulate breathing, but pinching the tail is a more powerful stimulant. After some time they will breathe regularly and turn bright pink; they will then be very active and will often squeal. At this stage, and not before, they are ready to be given to the foster dam. Resuscitation of a difficult litter may sometimes take an hour or more.

The foster dam is chosen from those in the clean colony that have had a litter 1–3 days previously: preferably about 48 hours. When the foster pups are ready, the dam is taken from her nest and placed in a spare cage. Her own pups are removed, and the foster pups put in their place in the nest, being well scented with bedding. The dam is replaced. If the pups are vigorous, and the operation has been properly carried out, the foster dam will care for and raise her fosterlings normally. She is, however, almost certain to reject weak, inactive or premature pups—as she would if they were her own.

It is a wise precaution to count the foster pups, not only at the time of fostering but on one or more subsequent occasions within the next few days. If the number increases, either some of the dam's own pups were accidentally left in the nest or she has given birth belatedly after fostering. In either event the litter has to be destroyed. The operation of hysterectomy has been described in some detail because it is a basic technique in breeding colonies of rats, mice and other laboratory animals. It is not difficult, but certain details are vital. In skilled hands it is entirely humane.

Variations of the above technique, which has particular reference to rats and mice, are used for a wide variety of other animals.

The use of hysterectomy and fostering, as a routine in large-scale breeding, will be described in Chapter 10.

CHAPTER 6

The Physical Environment

The laboratory animal is the material used by the experimental biologist for testing hypotheses and examining the effects resulting from the administration of substances. The selection of animal species and the determination of the degree of quality required are dependent on the nature of the experiments to be performed. Because of the financial considerations the operational policy of an animal division results from a critical analysis of the experimental purposes for which the animals are to be used. The success of that policy will depend to a very great extent on the nature of the physical environment in which the animals are to be bred and maintained.

The present chapter deals with the physical environment of the experimental animal—buildings, caging, ventilation and ancillary equipment—and the pests that may be encountered in animal houses. It is certainly not the intention here to present an ideal solution, which can cover all situations, but to outline basic schemes and variations. Every laboratory will have requirements and a physical situation which will differ from another, and decisions must be made in this context. Since the initial cost of providing animals facilities or modifications to existing premises is extremely high, it is important that a long-term view be taken when plans are formulated. Flexibility in design will therefore be stressed.

I. Animal Facilities

The animal species used for experimental purposes are, in the wild, extremely adaptable and in captivity can tolerate wide variations in temperature, humidity and caging. However, there are optimum conditions under which the animal thrives from the aspects of health and breeding performance. The laboratory animals that are used today have been, for the most part, selected and bred for many generations under laboratory conditions and consequently demand a high degree of environmental control for the maintenance of health. Species which do not have a laboratory-bred ancestry, such as cats and dogs, also require close attention to environmental conditions if satisfactory breeding results and standards of health are to be achieved.

Animal facilities can be considered in the light of three main functions—breeding, holding and experimental. Ideally each function should be kept

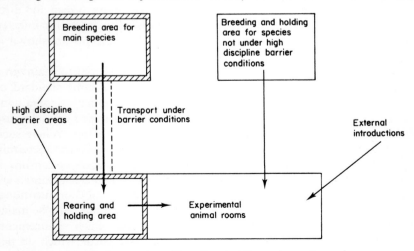

Fig. 6.1. Ideal separation of breeding, holding and experimental areas.

physically apart. The main source of infection in a well-designed and maintained animal house lies in the movement of materials and personnel. Separation of these functions should be complete, preferably with breeding facilities at a physically isolated site. This ideal situation, rarely possible in practice, is illustrated in Fig. 6.1. The breeding unit can be isolated by having its own staff who do not pass into other sections, and material access only through an autoclave and dip-tank. The holding unit and experimental rooms are subject to the movement of laboratory and animal staff, and frequently are subject to the introduction of animals from external sources and are concerned with a variety of species. Where all these functions have to be carried out within the same area, although rooms may be separate, it can be assumed that even the most rigid precautions will not prevent cross-contamination. The picture is not so black where only one or two species are bred and experimental usage confined solely to the output from such breeding. Where the usage is mainly from internally bred animals but some external introductions are inevitable, the latter must be isolated where possible from the former.

Three types of animal facility therefore exist.

1. The self-contained animal house.
   All animals used are bred internally or purchased from a single reliable source.
2. The partially self-contained animal house.
   Occasional external introductions.
3. The mixed animal house.
   Variety of species used and frequent external introductions.

The movement of staff and materials must be organized and the animal house designed according to which category is operative. Clean and dirty traffic paths must not cross, otherwise the possibility exists for infected material being conveyed to uninfected areas. These concepts are shown in diagrammatic form in Fig. 6.2.

In considering the design of animal houses it is essential to maintain a constant environment and discipline for animals of the same standard of quality. To breed or purchase pathogen-free rodents (under stringent conditions) and then hold them for experimental use in a mixed animal house without adequate precautions is a waste of money and effort. Where such animals are to be used for acute experiments, and daily consignments within the correct age and weight distribution can be made, special precautions in the non-breeding animal house need not be taken. Where such animals are to be used for long-term experimentation, as in toxicological or nutritional studies, then the same stringent environment and discipline must be maintained in experimental rooms as in the breeding unit. Such requirements can often be satisfied within the same building as animals which do not originate from carefully controlled breeding units and such an arrangement

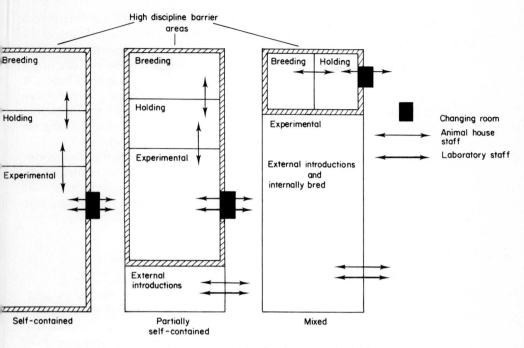

Fig. 6.2. Diagrammatic representation of main types of animal house.

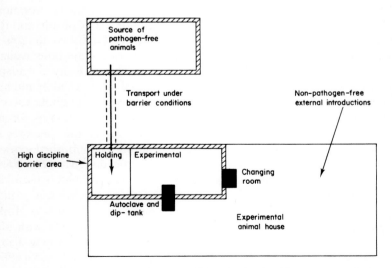

Fig. 6.3. Barrier area within mixed animal house.

is illustrated in Fig. 6.3. Under these circumstances it is advisable to permit entry only to laboratory staff who do not have to enter other parts of the animal house; these restrictions should also apply to animal technicians responsible for routine husbandry, where this is justified by the size of the operation. The training of animal technicians is dealt with in Chapter 8, but it should be mentioned here that in many instances the animal house staff can perform the routine functions associated with both the husbandry and experimental aspects of long-term programmes, thereby reducing the number of staff that have to enter these experimental rooms.

Considerations of layout are often dealt with in too superficial a manner, while the energies of those responsible for planning are concentrated on the details associated with construction and facilities. Layout decisions should result from an analysis of purpose and be modified to suit the physical conditions available.

II. Animal House Design

The provision of accommodation for animals should not be an afterthought; space which is unsuitable for laboratory purposes should not be automatically relegated for the housing of animals. Once animal accommodation has been constructed, the result is more inflexible than laboratory space due to the environmental facilities required. In consequence it is rational to pay greater attention to the design of animal than laboratory space. The latter, having greater flexibility with regard to purpose, requires less detailed planning and forethought.

A wealth of published information exists relating to the environmental and nutritional requirements of laboratory animals. Considerations of animal house design are sparse in the literature. Published designs are usually specific to particular purposes and the physical conditions available and are not generally interpretable for others. The necessity for adaptation within specific prevailing conditions is stressed in this chapter and basic principles form the main theme. The intending designer of animal accommodation is urged to visit a wide variety of existing animal houses, recently constructed, and formulate his own plans according to the successes and failures he encounters. There are, however, a number of published works to which the reader can be referred. The section on design in *The UFAW Handbook on the Care and Management of Laboratory Animals* is useful and the publication by Laboratory Animals Ltd of the first Laboratory Animal Handbook (1968) entitled *The Design and Function of Laboratory Animal Houses* is an up-to-date consideration of the details concerned with planning, construction and equipping. The following references may also be consulted—Hill (1963); Ottewill (1968); Nuffield Foundation (1961); Report (1969) (in Swedish) and Report (1963).

A. LAYOUT

It has been shown that three types of animal house exist: self-contained, partially self-contained and mixed. The way in which the layout is planned will depend on the type required, or the way in which they are related if a large facility is proposed. Whichever type or combination is planned four main subdivisions can be considered for each:

1. Breeding facility.
2. Experimental rooms.
3. Service areas.
4. Staff rest room and changing rooms.

1. Breeding Facility

As has already been discussed, this should be physically separated from the experimental areas and animals which leave should never be re-introduced.

2. Experimental Rooms

For the prevention of cross-infection it is advisable to design rooms which are small, even though the cost be increased. Where small rodents are used for extended periods of weeks or months and the experimental procedures restricted to simple operations such as inoculation, dietary variations or extraction of body fluids, rooms of size 6 m × 3 m or 4·5 m × 4·5 m provide adequate movement for staff as well as space for caging. Larger rooms with mobile racking may be used by many experimenters with consequent difficulties when all wish to examine or treat their animals at the same time. The ideal is for each experimenter to have his own room; although this can rarely be achieved, the provision of a larger number of smaller rooms will be more satisfactory than relying on a few large areas. In addition to this argument, larger rooms tend to be noisy and more susceptible to regions of unsatisfactory temperature and ventilation conditions. If infection breaks out in a large room, it is more difficult to control and more experiments will be affected. These latter arguments also apply to breeding rooms.

The experimental animal room should be regarded as an extension of the laboratory. Animals should not be removed to a laboratory for inoculation or examination and then returned. If this is done the chances of infection are increased and the health of the animal will suffer from the stress caused by transport and the change in environmental conditions. Provision should be made in experimental rooms for benching and apparatus, and electricity, gas and water made available in suitable positions. The movement of animals to laboratories should be restricted to acute experimentation only. For larger animals provision should be made for a room in which surgical operations with subsequent recovery are carried out. In such cases special provision should be made for housing during the recovery period away from other animals.

The incorporation of a quarantine facility is often overlooked but is nevertheless vital if the introduction of animals from external sources is envisaged. Such a room can be used as an isolation facility for long-term experiments where a disease outbreak has occurred or for isolating larger animals suffering from temporary but infectious complaints.

The inclusion of a soundproof room or rooms should be considered at the planning stage based on a long-term view of possible experimental requirements.

3. Service Areas

These should include the following:

a. *Chief technician's office.* To house all animal-house records and provide facilities for routine and occasional laboratory work connected with breeding and maintenance of the animals under his care.

b. *Storage.* 1. Storage facilities for food and bedding before sterilization. 2. Separate facilities for holding all sterilized material and equipment. 3. Storage for caging and racking not being used. 4. Storage for household items such as cleaning equipment, food containers, towels, sacks and so forth.

c. *Sterilizing* and *washing.* Where a strict barrier system is to be employed, the sterilizing equipment—autoclave or ethylene oxide sterilizer—will be sited between storage rooms on the clean and dirty sides of the barrier and will be double ended. For units not operating a barrier system the sterilizer should be sited in such a way that the storage of clean and dirty materials and equipment is kept as far apart as possible. It is advantageous that materials should be sterilized immediately before requirement in an animal

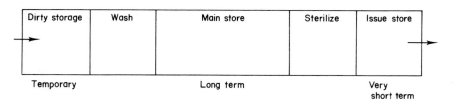

Fig. 6.4. Cleaning and sterilizing of cages and equipment.

room, or with as short a storage period as can be arranged within the operational routines of the animal house. Similarly, pre-sterilizing cage and equipment washing should be undertaken as soon after leaving the animal room as possible. This is shown diagrammatically in Fig. 6.4. Where disease is a major problem within a mixed animal house it is often advisable to sterilize equipment after leaving the animal room before going into storage and again before reissue. It should be remembered that where storage facilities are inadequate in a "conventional" animal house, the process of sterilization becomes more of a gesture than a function producing a worthwhile result.

4. Staff Rest Room and Changing Rooms

The provision of a staff rest room, where mid-morning and mid-afternoon breaks can be taken, is necessary to maintain good working conditions for the staff and provides a meeting point where problems can be aired and views exchanged. Changing rooms with showers are not only vital to the successful operation of a barrier unit but should be included in all animal houses since it affects the morale of the staff by supplying the requirements for personal cleanliness. The siting of these facilities depends on the type of animal house under consideration, but in principle should be situated as near to the personnel entrance as possible and, ideally, should form an entrance barrier to the animal house.

The functions of waste disposal, food preparation and post-mortem examination should also be considered.

5. Disposal Area

This area is concerned with the collection and disposal of waste materials such as bedding, nesting material, excreta and cadavers. Collection of these materials from the animals rooms should be by means of sealed bags, paper or plastic, in order to reduce the transmission of disease and to maintain a high degree of cleanliness. It is often not practicable to site an incinerator in the vicinity of the animal house and this necessitates the storage of waste until the time of collection. Small cadavers may be conveniently stored in plastic bags until incineration, but larger cadavers may have to be disposed of by other means if adequate incineration facilities are not available. In these cases it is useful to install a suitable-sized maceration unit, although careful siting is required to reduce the effects of the noise and vibration produced.

6. Food Preparation

This may not be necessary when only small rodents are utilized, since the incorporation of test substances into the diet is usually carried out in the laboratory because of the small bulk involved. Where large numbers of dogs, cats or primates are involved, or under conditions where complete diets are not utilized, a food preparation room is essential.

7. Post-Mortem Room

Where detailed examination, for experimental purposes, of large numbers of cadavers is periodically undertaken it is often found to be more convenient to site a post-mortem room in the animal house precincts. This room can be regarded as potential animal accommodation and should therefore be equipped with the necessary services.

Basic layouts are illustrated for the three main types of animal house. These are intended to illustrate basic principles and not inflexible plans, modifications depending on the nature of the experimental work and the limitations of existing buildings (Figs. 6.5, 6.6, 6.7).

To maintain a disease-free animal house complex certain layout principles must be incorporated.

a. Entry paths for materials, equipment and personnel into animal rooms should be distinct and not cross over exit paths. Where single corridors only are possible, exit material must be bagged and sealed and removed by trolleys restricted to that purpose. Staff outer clothing must be changed before entry into each animal room.

b. Laboratory staff entry is restricted as far as possible and only after a clothing-change routine and wash.

c. Store rooms and service areas positioned near an external access point. Sterilizing and cage-washing facilities positioned between dirty and clean stores.

It will be noticed from these outline plans that the storage areas are given a large share of the available space. While the absolute area is dependent on the intended use of the animal house and discipline to be imposed, it is a common failing that the provision of storage space is inadequate. Nothing is more conducive to a deterioration in operating discipline and sloppy housekeeping, than a situation where experimental rooms, or even corridors, have to be used as storage areas for caging and equipment. Inadequate food and bedding storage facilities may result in the overflow being housed in areas totally unsuited for the purpose.

The proportion of space allocated to service areas increases with the degree of discipline imposed—from the "conventional" unit where only minimal precautions are taken, to the strict barrier-maintained unit. In mixed or partially self-contained units, it will be necessary to duplicate many of the service areas in order to reduce the possibility of cross-contamination.

B. CONSTRUCTION

The range of available materials used in the construction of animal houses is vast. Some have proved to be successful, some not. To a great extent the planner will be in the hands of the architect and builder, who may have their personal preferences; it is the responsibility of the laboratory head to ensure that the basic criteria are decided and all information relating to purpose, environmental requirements and equipment to be used are conveyed to the architect or builder. His ability to communicate this information will depend on the thoroughness with which the planning group have produced their brief. The alteration of plans during construction—even

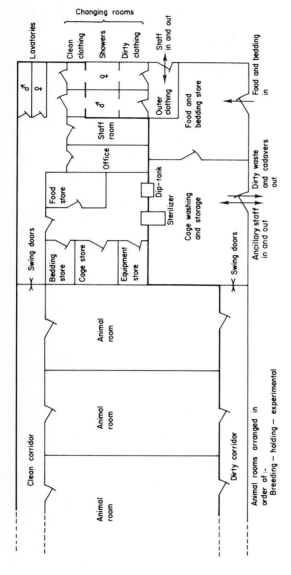

Fig. 6.5. Schematic layout of self-contained animal house.

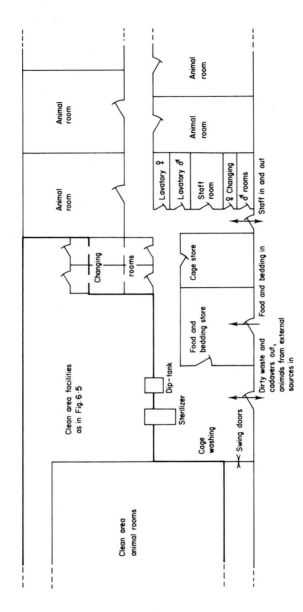

Fig. 6.6. Schematic layout of partially self-contained animal house.

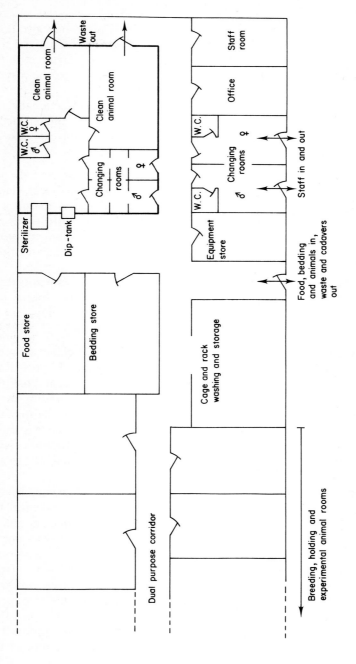

Fig. 6.7. Schematic layout of mixed animal house.

relating to apparently trivial items—can be costly and cause undue delays. The amount of capital available may require alterations to the brief and it is advisable to consult the architect during the early planning stage. It is helpful to incorporate requirement limits, where they can be set, and to apply priority ratings to the various aspects; this enables the architect to provide a provisional plan without constant back reference because the cost exceeds the stipulated limit.

The work of the planning group is essential to the functionally satisfactory and economically reasonable construction of an animal facility. In order to save the time of personnel involved, the team can exist as two bodies.

1. The executive group: comprising the laboratory director, the administrative head of the animal facility, and chief animal technician. They will be concerned with the formulation of the brief, liaison with the architect or builder and the Home Office, making decisions relating to alterations in the brief and confirmation of the architect's proposals. They will also be responsible for the final acceptance of the completed project.
2. The planning group: comprising the executive group together with departmental heads, project leaders, those requiring special facilities or having relevant specialist knowledge and the competent person concerned with the use of radioactive substances.

This group is concerned with providing all data related to species of animals to be used, quantities, quality of animal required, types of caging, husbandry routines, environmental requirements, special facilities and expected future developments. These data will then be used by the executive group to determine, in conjunction with the architect, the animal house layout, finishes and equipment to be installed. At this stage it is also necessary to determine the staffing structure of the animal house and to lay down the routines that will be employed. This will enable the architect to appreciate the entire operation and plan accordingly.

Having defined the requirements and a tentative plan having been made, it is then necessary to consult the Home Office inspector to ensure that the proposals do not conflict with current acceptable practice and guide lines laid down. Although the Cruelty to Animals Act of 1876 does not specifically cover areas used for breeding laboratory animals, it is nevertheless advisable and courteous to include these areas since recommendations by the Littlewood Committee (Report, 1965) contain proposals for their inclusion; in any event, the inspectorate have wide experience gained from first-hand knowledge of many animal facilities and can often provide valuable advice. It is mandatory that experimental facilities be approved by the Home Office before use and therefore sensible to invite comments early in the planning stage when definite proposals can be outlined.

A great deal of valuable help can be obtained by visits to other laboratories and so learning by the mistakes of others. It is fortunate that in this field information is very freely exchanged and errors are pointed out with the same enthusiasm as successes.

The importance attached to the preparation of the brief by the planning group cannot be overstressed. An ill-prepared, incomplete and superficial direction to the architect will produce constant reference back for clarification and increases the chance of costly mistakes due to misunderstandings. The meetings of the planning and executive groups must be minuted and arranged on a regular basis and a secretary appointed to collate the data and prepare the brief.

A scheme of progress in the planning, construction and approval of animal accommodation is given in Fig. 6.8.

1. The Building

It is often said that where animal accommodation is concerned, successful operation depends on the amount of money the user is prepared to invest. While in extreme cases this is possibly so, the success of an operation depends to a greater extent on careful planning, and by keeping the basic requirements in mind this need not depend on a costly method of construction. Where separate buildings are being considered, the use of prefabricated buildings is often satisfactory for many purposes and insulated timber buildings can be used for small rodent breeding. The advantages of the latter are that they can be arranged to give separation between the animal rooms and may be linked by corridors which are open to the environment, thus reducing the possibility of cross-infection. Timber buildings provide greater flexibility for future modifications than brick or pre-cast structures and they are cheaper. Modern finishing materials enable them to be washed down internally if required. Modifications to existing animal houses, or conversion of an area within an existing building, should be carried out using materials which will not affect future flexibility. Partition walls between animal rooms can be constructed of lightweight partitioning if no shelving or cage rack brackets are to be fitted, provided the internal finish allows for complete sealing of each room. The range of materials for constructing free-standing buildings is very great and the choice depends on the purpose, the site and the design. The most important and universal considerations are related to insulating properties, ease of installation of vermin barriers and the working conditions for staff.

The approaches to external buildings must be designed and constructed at the same time as the main building. It must be remembered that food and bedding are often delivered by articulated vehicles and turning room will be required. The road construction must be able to stand this type of heavy vehicle. The siting of a car park for staff should be away from delivery and collection areas.

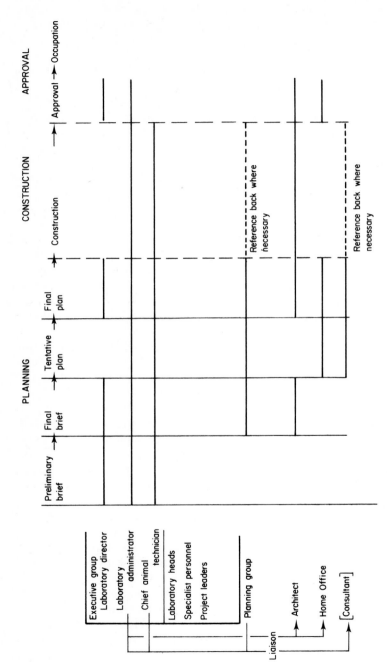

Fig. 6.8. Responsibilities in planning, construction and approval of animal accommodation.

Whatever the design or materials used, all floors should be on the same level without any steps, since most movement in an animal house takes place on wheels. Where slight changes of level are necessary, a gentle ramp should be used.

2. Floors and Walls

The tendency in modern animal houses is to do away with washing down walls and floors in animal rooms. The advantages of this procedure have always been in doubt and its origin lies possibly in a psychological desire to wash away uncleanliness rather than in scientific principle. The exception is in dog accommodation, where the only satisfactory method of removing faecal material from pens and runs is by high-pressure hose. Hosing down in an animal room produces high humidity, unpleasant working conditions, probably distributes infection rather than eliminates it and produces problems associated with floor drains. It is surprising how difficult it can be to lay a floor which is graded towards a drain; either the slope is too great, producing mobile rack instability, or pockets of water remain after hosing. The drain traps tend to get clogged easily from sawdust and are a perpetual worry as a source of infection. The process of hosing as against mopping also limits the choice of flooring material. Many animal houses now operate a "dry" routine consisting of suction cleaning of cages or use of paper over trays where wire-mesh bottom cages are used: floor cleaning by use of industial vacuum cleaners; or simply a daily sweeping.

By operating a system of room rotation in breeding units or total evacuation at the end of an experiment, rooms are periodically sealed and fumigated. It is an axiom to state that the animal cage and immediate environment are as clean as the animal; it would appear to be a waste of effort to try and sterilize walls and floors when the source of infection remains untouched. Where animals are free from infection and are replaced from reliable sources, the dry system of husbandry works extremely well. Where infection exists and is inherent in the animal house, hosing down will do little to affect the condition.

The abandonment of hosing down provides the planner with a wider choice of flooring and wall finishes. These are described by Statham (1968).

Where animal cages are held on wall racks or mobile racks are not very heavy, polyvinyl chloride welded sheets make a serviceable floor which is easily swept or vacuum cleaned or mopped, provides warmth underfoot, is attractive in appearance and does not crack. Tiles should be avoided since the joints provide areas where bacteria will accumulate. The final choice will depend on the nature of the undersurface, the method of husbandry and the method of caging. The working conditions for staff should not be forgotten and the floor should be slip-proof and not detract from the appearance of the animal room.

Wall surfaces should be free from rough surfaces, waterproof and have a sealed joint with floor and ceiling. Tiling and glazed brick have disadvantages in that they are susceptible to cracking and the jointing is not permanently impervious to moisture. A plaster finish painted with a plastic paint is suitable or an epoxy-resin paint sprayed on to the brick surface is durable and attractive and requires little maintenance. Vinyl sheeting can also be used where free-standing caging racks are used. The choice of colours should be imaginative and consideration given to different colours for different rooms rather than a uniform shade throughout. Although rarely visited by those responsible for the planning, the animal house is the constant working environment for many people—weekdays and weekends—and consideration should be given to making that environment as cheerful and pleasant as possible.

Since most movement in the animal house is on wheels and doors tend to suffer from frequent knocks and abrasions, it is advisable to apply metal cladding to the lower half. Glass panels should be inserted in the upper half to avoid injury on opening and to avoid unnecessary entry when a staff member is being sought. Doors should be flush-panelled for easy cleaning. Rooms where a footwear change or fitting of overshoes is a part of the discipline can be fitted with a movable length of timber about 23 cm high across the floor of the door opening; this acts as an indication and reminder of the discipline required and is simple and cheap to fit.

Windows play an essential part in providing acceptable working conditions in all animal and service rooms except where experimental requirements or the restrictions of the existing building prevent it. The maintenance of uninterrupted breeding in most species depends on the extension of the photoperiod during the winter, rather than its reduction during the summer, therefore windows should be provided where they can improve the working environment. However, it should be borne in mind that windows should not overlook areas where casual passers-by, not associated with the laboratory, can see into animal rooms.

3. Ceilings

The temptation to use the areas near the ceiling edges for running services such as water pipes, electrical conduits and ventilation trunking should be avoided. This practice produces cleaning problems and they remain a constant source of infection where bacteria can collect. The ceiling should be as free from suspended services as possible. Where a mezzanine floor exists, carrying services, a suspended ceiling can be used provided it is completely sealed to the walls using a plastic material such as "Plastapak". An unsealed suspended ceiling with services above, which can only be reached from the animal room, must be avoided since the service area will create a source of dust and infection. The materials of the ceiling should be chosen bearing in mind the requirement of fumigation under conditions requiring air saturation.

4. Services

Where possible, and this should apply in the majority of cases, main services should be sited in corridors with the minimum of exposed ducting and conduit into the animal rooms. Sinks and cupboards should be kept to the minimum and should be designed and sited to provide for easy cleaning and the elimination of areas where dust or moisture can collect.

All electric fittings must be of the waterproof type. Even if cleaning is restricted to dry methods with occasional mopping with disinfectant, it is necessary during fumigation, for the elimination of nematode eggs, for this process to be carried out in a saturated atmosphere. Electric systems, such as time switches for controlling the photoperiod, should be sited outside the animal room in the corridor.

Conduit for electric points, gas and water pipes should be set into floors and walls for easy cleaning. Apart from a sink, each animal room should possess a water point for the attachment of a hose, which may be required for a device for filling or cleaning water bottles. The installation of several electric socket outlets, particularly in experimental animal rooms, although not immediately required, will provide for flexibility if the purpose of the room alters. In any event two outlets should be regarded as the minimum for use with balances, hair clippers or cleaners, whatever the purpose of the room. Gas outlets should also be fitted with flexibility in mind.

The fitting of small-diameter copper tubes passing from each room to the corridor is useful to enable a quick check to be made of the pressure differential.

C. FLEXIBILITY

The process of planning and constructing animal facilities and, in some cases, gutting existing facilities is extremely costly and demands a great deal of time from senior laboratory staff. While the construction is in progress delays in the progress of research projects may result. Although it is not easy to visualize the demands of research several years in advance, an attempt should be made at the planning stage and these factors taken into account. It is better to err on the side of generosity than be faced with another major upheaval in a few years' time. Animal house accommodation can be designed on too narrow a concept; mention to date has been of "animal rooms" and not "rat rooms" or "guinea-pig" rooms. This has been deliberate. The equipment for holding and breeding laboratory animals which is now available is flexible in its environmental requirements. The use of adjustable wall racks which can take a variety of cage sizes provides for flexibility in animal rooms where racks are confined to areas adjacent to walls. Mobile racks provide flexibility but produce storage problems when not in use; unfortunately, the universal rack to take all cage sizes and types has not yet been designed. However, many modern techniques, such as tray breeding of guinea-pigs, enable animal rooms to be used for a variety of purposes.

Certain species require special accommodation which cannot easily be utilized for other purposes. The housing of dogs is a particular instance of this inflexibility and such accommodation should be considered very carefully with a view to the future. Breeding accommodation for certain species such as cats, rabbits and dogs require to be tailor-made if the operation is to be successful and economic. Because of this inherent inflexibility very careful consideration of long-term requirements is necessary before planning for such activities.

Provision should always be made for additional animal rooms than would appear to be immediately required. Even if demand remains static on average there will be peaks of activity when the demand for space cannot be satisfied if the design has been trimmed to an average requirement. Additional space is extremely useful as a quarantine area, the provision of continuous accommodation when rooms are routinely fumigated and also, if fitted out to such a standard, temporary laboratory accommodation.

The provision of flexibility in terms of temperature control and ventilation will be discussed in a later section, but it should be remembered here that these factors are essential for the successful use of multi-species animal rooms.

D. PROVISION FOR LARGER ANIMALS

Many of the larger animals used in biological research do not adapt well to conventional accommodation in the animal house and special facilities must be provided. Notable among these species are cats, dogs and primates. It is necessary to differentiate between short-term holding and breeding facilities; while satisfactory holding facilities can usually be provided at a low cost the provision of breeding facilities requires special consideration. It is not intended to describe such facilities in detail since many excellent publications exist on the subject (see Bibliography), but rather to point out the general nature of the special requirements and to highlight the difficulties which may be encountered.

1. Dogs

The most widely used breed of dog for experimental purposes is the beagle. This choice has been made on account of its size, temperament, amenability to maintenance in packs, ease of breeding and the extent of published background physiological data and information on the incidence of abnormalities. Where the animal is to be used for acute experiments, it may be kept overnight in a large cage—the collapsible type of about 1m ×60 cm is useful—without a run facility. For longer holding periods or postoperative maintenance the accommodation must be specifically designed and an external exercise run included. The prime requisites are a dry kennel free from draughts, adequate ventilation, heating—preferably of the underfloor type—and constant access to drinking water. Plenty of fresh air will satisfy the need to keep cool in hot weather, but where this is not possible provision must be made for cooling the environment.

The dog accommodation should be sited away from other buildings, where possible, to provide the fresh-air exercise run, reduce the disturbance from barking and to keep the animals remote from the general public. Where this is not possible it should be sited at roof level. In designing accommodation for dogs four factors are of prime importance.

a. Constant access to runs will reduce soiling in the kennel and so simplify the cleaning task and reduce the problem of maintaining a dry kennel area. This should be a consideration whether the dogs are kept singly or in groups. During hot weather, this constant access will help the dogs to keep cool and will reduce staff time taken in moving dogs to a remote exercise area. Where accommodation has to be provided within the framework of an existing building or where space is limited this can rarely be achieved, unless the numbers to be held are very small.

b. The ventilation in enclosed areas should achieve five to ten air changes per hour. These figures imply ventilation in the kennel area where the dog is living and not measurements taken at ceiling level where the input may be sited. Kennels are often constructed of solid material which allows no air movement; under these circumstances the calculated air changes may be five to ten per hour for the room as a whole but may approach zero where the dog is situated. A damp dog is an unhealthy and uncomfortable dog; properly designed ventilation together with underfloor heating in the kennels (hot-water systems are a cheap and effective way of providing this) do much to provide a satisfactory environment. Since most of the cleaning of dog accommodation has to be accomplished by high- or low-pressure hoses, the problem of damp conditions must be kept constantly in mind during the design phase.

c. Faulty design and construction of drains are a frequent cause of serious dog management problems. The quantities of water used in cleaning are considerable and dog faeces have a high fat content which increases the chance of drain blockage. Since many units have to be sited in remote areas, sewage treatment may have to be carried out on site by cesspool construction. The size of such construction should relate to these factors. Falls in runs and kennels should be adequate to prevent the formation of damp areas, and gulleys of at least 15 cm width be provided to take the faeces and water to the drainage system. Too much attention cannot be paid to the method of cleaning and the drainage system, since bad design is extremely difficult to rectify and leads to unhealthy animals and bad working conditions for staff.

d. Dogs bark. This noise can be severely disturbing to staff during the day and may cause a nuisance to residential areas at night, even if sited some distance away. Dogs which are bred and raised in laboratory kennels are usually quiet if undisturbed, but if one dog starts the others usually follow suit. The way in which this can be overcome will depend on the siting of the unit in relation to existing high ground and the direction and distance of residential areas; in any event the advice of an architect experienced in sound containment should be sought. For the internal reduction of noise, walls and ceilings can be soundproofed and the dog accommodation shielded from external noise disturbance by siting service areas and corridors around the kennels and pens.

For breeding purposes it is sufficient to allow for larger kennels of about 1·8 m × 1 m to accommodate pregnant and lactating bitches and for rearing young puppies. Again, the facility for constant run access allows the bitch to train her pups and reduces fouling of the kennel area. A movable littering box made out of stout timber is necessary and a platform situated about 50 cm from the floor allows the bitch to escape from her puppies.

The best advice to those about to design dog accommodation is to visit other laboratories and learn by their successes and mistakes. The expertise lies in interpreting these data in terms of the specific conditions that will have to be met in the new unit.

2. Cats

Cats may be held in animal rooms, provided the ventilation is adequate, for long periods without detriment to their health. Accommodation as a group is economical in space but produces cleaning problems and fighting can break out. A convenient method of holding animals is to hold the cats in pairs in collapsible mesh cages of about 1 m × 60 cm in size, placed on wall racks or trolleys. The cat is fundamentally a very clean animal and the provision of a sawdust tray will satisfy the need for cleaning. Ventilation under these conditions is of paramount importance and should be in the order of ten to twenty air changes per hour. The temperature should be maintained at 18°C and should not vary by many degrees above or below this figure. High temperatures are tolerated less well than cool conditions. Respiratory infection is the main problem to be overcome, and correct ventilation and temperature are of prime importance. The provision of cardboard boxes in the cages allows the cat to escape from draughts and provides sleeping accommodation.

Accommodation for cat breeding is still in the experimental stage. Of all common laboratory species the cat presents the greatest challenge in the production of healthy animals at an economic cost. The most successful units have been developed from the "Pagoda" system of W. L. Weipers (Scott, P. P., 1967) and are in operation at Smith, Kline and French Laboratories Ltd (Quinn, E. H. and Pearson, A. E. G., 1968) and Roche Products Ltd, both in Welwyn Garden City, Hertfordshire. This system relies on the principle of ventilation by constant access to fresh air, together with a "hot spot" provided in the hutch section by a tubular heater. Mating runs are provided where up to twenty-five queens are accommodated with one or more breeding toms, the latter being rotated at intervals of 4—7 days.

The construction of free-standing accommodation presents far less of a problem for cats than for dogs. The buildings can be of timber, which allows for flexibility in design and permits economical construction of specialized internal accommodation. Soiling trays satisfy the disposal of excreta and so eliminate drainage difficulties. Runs may be constructed with concrete or rolled ash floors, but overhead netting is necessary and electrified perimeter fencing may be required to prevent contact with wild foxes or dogs. External runs should be partly covered by corrugated plastic roofing as a protection against rain. "Pagoda" units must be fly-screened and it is advisable to carry this out for external runs as well.

3. Primates

The housing of small numbers of quarantined primates presents little problem in a well-designed animal house. The cages can be accommodated in a conventional animal room provided flexibility has been incorporated into the overall design, so that the required levels of ventilation and temperature can be achieved. However, there are special points which should be considered before this is undertaken. Primate cages and their occupants are heavy and it is advisable that they be supported on two-tier free-standing units. If cantilever wall racking is to be used the advice of an architect or builder should be sought to ensure that the strength of the walls is adequate for the purpose. Primates indulge in vigorous cage shaking and this sets up considerable vibrations which further add to the stress applied to the racks. This habit also produces a high noise level, and if the room cannot be sited away from other animal rooms it should be suitably soundproofed.

Ventilation should be at the rate of twelve to fifteen air changes per hour where humidity can be controlled at 50—70 per cent R.H. and a temperature of 18—20°C for macaques maintained, young animals requiring a minimum of 22°C. Care should be taken not to overcrowd a room and a minimum distance of 2 m between cage rows is necessary for ease of handling and prevention of injury to staff by the extension of arms through the cage fronts. Windows should be wired to prevent escape and doors locked after entry and exit if a double-door system is not practicable.

Under circumstances where large numbers of monkeys are to be maintained, unquarantined animals introduced, breeding facilities required, or experimentation undertaken involving human pathogens, a specially designed self-contained unit must be constructed. The reader is referred to Coid, C. R. (1968) and *The UFAW Handbook on the Care and Management of Laboratory Animals* for detailed information in these cases.

Wherever primates are to be housed it should be borne in mind that both staff and animals are at risk from each other in the passing of infectious disease. Outer clothing changes, the wearing of face masks and scrupulous hygiene are therefore mandatory. Provision of a second primate room is advisable for the isolation of sick animals from healthy stock.

4. Other Species

The housing of rabbits, guinea-pigs, hamsters and other smaller species can again be accomplished within the well-designed animal house in standard rooms. The ventilation and heating system must, however, have the required flexibility. While metal cages are probably still the most satisfactory for housing rabbits, guinea-pigs can be accommodated in groups in fibreglass trays on multi-tier racks and smaller rodents such as hamsters can be housed in standard rat cages. For the requirements of the less common species the reader is referred to *The UFAW Handbook on the Care and Management of Laboratory Animals*. Given imagination, ingenuity and an animal house designed for flexibility, much can be accomplished in providing suitable accommodation for these species.

Where breeding facilities are required, separate units may have to be built if considerable numbers are to be produced. The exception is the breeding of guinea-pigs. The development of the technique involving the use of fibreglass plastic trays on mobile racks enables about 1200 animals to be produced per year in a room size of 6 m × 6 m. The system is based on one boar running with five sows in a 1·2 m² tray together with progeny up to 3–4 weeks of age. The same trays are used for growing and holding stock. The younger the animals are used the more space becomes available for breeding groups.

The economical breeding of rabbits is best accomplished by the use of commercial breeding batteries, housed in a separate building. Similarly, where large numbers of poultry have to be accommodated the commercial battery unit provides a cheap and labour-saving solution where sufficient space is available.

E. PROVISION FOR THE USE OF HAZARDOUS MATERIALS

1. Radioactive Materials

Where tracer studies only are carried out the provision of floor drainage is the only essential additional feature to be incorporated in the animal room. This drain may be connected to the sewage system provided this is agreed by the local authority responsible for licensing the premises for such work. The granting of this permission will depend on the quantity and nature of the radioactive material permitted in the licence. As in all animal rooms the provision of washing facilities must be incorporated.

Where high levels of radioactive material, particularly from gamma-emitting nuclides, are to be used, it will be necessary to design a self-contained unit for the purpose. Such a unit will require special features for the protection of personnel and the environment as well as containing the features of a well-designed conventional animal house already discussed. Such a unit is best sited as a free-standing unit since it should incorporate facilities for the reception, presentation and storage of the radioactive material, and the post-administration laboratory work should be undertaken in the same area. This will involve the construction of sophisticated changing facilities, fume hoods, filtered extract ventilation, sealed drainage and disposal facilities. For a detailed discussion of these requirements the reader should refer to Hughes (1968).

2. Infective Agents

This topic will be discussed in detail in Chapter 8, where human pathogens are included, in relation to the safety of staff. The problem is basically related to containing infection in the vicinity of the affected animals. The design of animal rooms is related to keeping a healthy animal healthy by preventing outside contamination from reaching its vicinity. The animal contaminated with highly infective pathogenic organisms receives the reverse treatment.

Large experimental animals, particularly farm animals and *Equidae*, require specially designed accommodation remote from other animals—or buildings, where the infective agent may be pathogenic to man. The facilities required will include separate loose-boxes for each animal, constructed with smooth internal surfaces and rounded edges, so that cleaning can be effective. A ventilation system must be designed to provide a negative pressure within the loose-boxes and air is filtered before entry. Extracted air must be filtered to remove all infective organisms. Personnel access is restricted to the minimum required for husbandry and experimentation. The reader is referred to Seller (1968) for details of the design and construction of large animal units for these purposes.

Small animals are best maintained in rigid isolators. Where organisms not pathogenic to man are used, it is possible to use a whole room treated as an isolator. The filtered air from the animal-house plenum system may be used, but extracted air must pass through a separate filtration system to prevent spread of infection to other parts of the animal house; the infected room must be maintained at a negative pressure in relation to corridors, and air-tight double-door access is advisable. All materials leaving the room, including bedding, faeces, cages, instruments and outer clothing, must be sterilized and conveyed to the sterilizing area by a double bagging method. Ideally, a double-door sterilizer should be incorporated into the design of such a room.

Where small numbers of animals are to be used, it may be more convenient and cheaper to confine them in rigid isolators. This method is safer than utilizing an entire room and is essential where human pathogens are involved. The rigid isolators may be fed with air from the main plenum system or by individual blowers. Intake air should be filtered and it is essential to pass the extracted air through an efficient filter to remove all contaminants; an extract fan must be fitted to enable the isolator to be maintained at a negative pressure in relation to the room. All food, bedding and instruments must be pre-sterilized and entry to the isolator made by means of a dunk tank or sterilizing lock. Removal of material from the isolator should be by the same route and technique with immediate subsequent sterilization. The material with which the isolator is constructed should be able to withstand autoclaving and this requirement will also govern the external dimensions. The design should conform to the general principles associated with germfree isolators, but exact details will relate to the nature of the work undertaken. Even though such equipment should prove virtually foolproof in operation, if properly designed and subjected to frequent checks, it should nevertheless be sited in an animal room situated in as remote a position as possible and not in a laboratory.

III. Animal House Equipment

The types of racks, cages and general equipment to be used in the animal house should be decided on at the time the first brief to the architect is prepared. Wall racks may require strengthened walls; the use of large trolleys may determine the width of corridors and doors and the facing materials used in their construction; the method of sterilizing and equipment to be used must be known before the service areas can be designed.

These decisions are not as involved or complex as might at first appear. The past decade has achieved great advances in the knowledge of the physical requirements of the laboratory animal, and extensive practical experience has been gained from the use of new materials and the employment of new ideas. Some have failed the test of time, some have proved advantageous. The temptation, as common as a decade ago, to order equipment to a non-standard specification to indulge a personal whim or prejudice, is not now so frequently encountered. Special equipment made to order is far more costly than standard items; because the latter are cheaper does not mean a reduction in effectiveness but reflects the economics of mass-production. Where the more commonly used laboratory animals have to be housed a decision on equipment is best made as a result of visits to other laboratories with similar requirements, when equipment in use can be examined and merits and faults can be discussed with those possessing practical experience. For many species, because their use is not widespread or numbers used are small, there is considerable variation in equipment used, resulting from inadequate long-term assessment of their needs. This variation may also stem from widely differing environmental conditions between one animal house and another. Under these circumstances intelligent analysis of the problems and experience of others in relation to the basic requirements of the animal must suffice for the basis of a decision. The material of which the equipment is constructed, as well as the design, is an important consideration. Cheap material may result in heavy costs due to short effective life; on the other hand, materials may be too durable with the resulting inflexibility if new innovations render the equipment obsolescent. For example, zinc was once hailed as the solution to the housing of small rodents in boxes, due to its resistance to gnawing and corrosion and the detrimental effects of sterilizing; there are many laboratories who still use such cages since they feel they cannot justify conversion to more satisfactory equipment which would result in the writing-off of large numbers of serviceable cages. Probably the solution is to aim at reasonably priced equipment having a life of about 5 years. This provides for a reasonable write-off period and ensures that new developments have been thoroughly tested before replacement is necessary.

The equipping of a new or modified animal house is a major cost item in the overall bill and decisions should be made only after the user is satisfied that all possibilities have been examined and the final recommendation represents the most satisfactory solution to the individual problems of the laboratory.

A. RACKS

Where fixed racking is to be employed, either as brackets from the walls or suspension from the ceiling, flexibility should be the keynote in respect of shelf width and distance between shelves. Provided the lowest shelf is not less than 30 cm and the highest not more than 1·5 m from the ground, the easiest way of satisfying this requirement with wall racks is to use slotted vertical bearers fixed to the wall, to which brackets of varying length can be fixed in the desired position. This will enable various sizes of cages to be accommodated at maximum capacity for each type. The advantages of wall racking are most apparent in the rectangular room of about 3 m width, as seen in Fig. 6.9.

The cage supports, to fix on the brackets, may be hollow rods or bars which are not fixed but lie in recesses in the brackets. This system provides for easy cleaning and storage of the rack components. The distances between the longitudinal brackets depend on the cage size to be accommodated. Solid shelves have been advocated to avoid the contamination of cages by material falling from those at a higher level. While this is probably advisable where wire-mesh cages are used, it is unlikely that where solid-bottom or grid-floor and tray cages are used the advantages will outweigh the disadvantages introduced from increased cleaning and storage requirements.

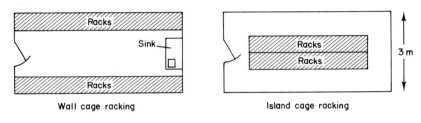

Fig. 6.9. Positioning of cage racking in 3 m wide room.

Racks suspended from the ceiling have the advantage of simplifying floor cleaning in large animal rooms where central accommodation is required. However, in these instances, mobile racks provide greater flexibility and comparable capacity. The disadvantage of equipping experimental animal rooms entirely with mobile racks lies in the need to have different racks for different cages. This can produce considerable storage problems for racks not in use. Where experimentation is expected to be predominantly with one species and one type of cage can be used for all experimental purposes, the use of mobile racks is convenient. Where a number of species are to be used or different cage designs for different purposes with a single species, then a proportion of the accommodation in the form of wall racking is to be preferred where possible, to reduce capital outlay and relieve the storage problem. If the demand is entirely for suspended caging then there is little choice, mobile racking must be used.

A convenient method of breeding or holding large numbers of small rodents in the small animal facility, where space is at a premium, is the battery unit. The disadvantages lie in the lack of ability to observe the animals without pulling out the individual cage and, because of the close-packed design, satisfactory ventilation is difficult to achieve.

B. CAGES

The key consideration in the selection of cages is simplicity; this provides ease of cleaning and stacking and reduces cost. Little need be said of the basic requirements for the common laboratory species, since it is assumed that the reader is unlikely to indulge in *de novo* design and commercially available equipment has been developed and tested over a number of years. The dimensions required depend on the species and purpose. For breeding purposes it is advantageous to examine the type of cage in use in a number of laboratories and discuss the merits and disadvantages with the users. For experimental purposes an attempt should be made to select the simplest cage design which conforms to the experimental requirement.

The principal factors to be considered in cage selection are:

a. *Ventilation.* Where variations in ventilation rate or temperature fluctuations are likely to occur, the solid cage is to be preferred. In wire-mesh cages the animal is subjected to a micro-climate identical with that of the room; in solid cages, and to a lesser extent with wire-mesh floor cages, the animal can affect its micro-climate by its own activities.

b. *Materials.* Wood is still occasionally used for caging small rodents, since it provides good thermal insulation, is cheap and can be sterilized by steam or boiling. However, such cages are difficult to clean and store and tend to vary in size, so making lid fitting difficult. Metallic cages have been made of zinc, galvanized steel, aluminium and stainless steel. Zinc is light and durable but expensive and not now used. Galvanized steel is nowadays mainly used for larger cages because of its strength, it is cheaper than most other materials but the galvanizing may be attacked by urine or liable to mechanical damage; it is also heavy and not amenable to a fine finish. Aluminium or aluminium alloy is light and durable but susceptible to gnawing. Stainless steel is exceptionally durable and can be used to produce cages with a polished finish with rounded edges which are easy to clean, but the cost is high and the durability may be a disadvantage—as already discussed. For caging small rodents (see Porter, Scott and Walker, 1970), plastic materials have proved extremely successful. They have the advantages of being light, easily cleaned (particularly by vacuum methods), stackable, reasonably resistant to frequent autoclaving and, by preventing an edge being presented to the animal, are resistant to gnawing. While the polypropylene cage is the most common and cheapest in use, the use of polycarbonate, although more expensive, is advantageous in that it is virtually transparent and the animals can be observed without being disturbed. All plastic cages require lids of a rigid mesh and stainless steel is almost universally used for this purpose. Where a mesh base is required for suspended cages, this is available in stainless steel with a plastic tray situated below. Fibreglass-reinforced resin is also used for larger cages such as guinea-pig trays or rabbit cages. More extensive use of this material in the animal house can be envisaged because of its durability, lightness and resistance to chemical and mechanical damage. Because of the smooth finish which can be achieved it is very easy to clean and is resistant to gnawing. It would appear to be an ideal material for individually caging larger animals such as cats and dogs in conjunction with anodized aluminium or stainless steel for doors and floors, particularly in constructing metabolism cages for these larger species.

c. *Maintenance.* Crevices should be avoided for easy cleaning. The material should be able to withstand frequent autoclaving. Durability should be considered in relation to a 5-year period.

d. *Storage.* Cages should be capable of being stacked for easy storage. Large mesh cages should be collapsible.

e. *Light.* Breeding cages for rabbits and ferrets should possess a dark hutch area; otherwise, if intense light is avoided, no design features relate specifically to this factor.

f. *Defaecation.* It should be remembered that most rodents, ferrets and dogs defaecate indiscriminately and therefore mesh floors are advisable for these species. Rabbits defaecate and urinate in selected corners, usually away from the source of food and can therefore be kept clean in a solid bottom cage. Cats will almost universally oblige by soiling only a provided tray containing an absorbent material.

g. *Washing and sterilizing.* In disease-free rodent breeding units, cages require only occasional sterilization, the routine cleaning and bedding replacement sufficing for the health of the animal; sterilization is undertaken on a room or section rota basis when the economic breeding life of the occupants is at an end. Cages used for experimental animals should be sterilized at the end of each experiment and returned to stock. The choice of cage will depend to some extent on the purpose in relation to frequency of sterilization. It is necessary to have to differentiate between sterilization and cleaning. The latter operation consists of manual or mechanical washing using a detergent and hot water with subsequent hot-water rinsing. Because a mechanical method can operate at high jet pressures and temperatures, the efficiency is very much greater than a manual operation. However, investment in a commercial machine is probably not justified by a turnover of less than 500 cages per week. The stress applied to the cage material is less with washing than with sterilizing and for most purposes a mechanically washed cage can be regarded as sterile. Where absolute sterility is paramount it is necessary to sterilize by prolonged immersion in disinfectant, autoclaving, steaming or irradiation. Of these methods, autoclaving is the simplest and, since this equipment should be standard in all animal houses for a variety of sterilizing duties, avoids duplication. It is essential that cages are selected which pack conveniently into a washing machine or sterilizer and will withstand the process selected.

h. *Size.* Some authors have proposed formulae to determine the optimum cage size for housing animals, depending on their individual weights and numbers. This can be of prime importance where space is limited and maximum density desired. Such a formula has been suggested by Kállai (1971):

$$\text{Area of cage in square centimetres} = (W^{0.75})(N^{0.85})$$

where W = weight in grammes of individuals and N = number of animals. Such formulae do not take into account such variables as the physiological or behavioural requirements of individual species, type of cage used or environmental conditions prevailing in the animal room—particularly the efficiency of the ventilation system. They must, therefore, be regarded as of academic rather than practical value.

The experience of others must be considered as the best guide, having due regard to the variables mentioned above. For uncommon species the only solution in determining optimum cage size is experimentation over a long period.

i. *Labelling.* Provision should be made on the front of the cage for easy label insertion and withdrawal. Many excellent plastic cages do not have this facility, and much time and nervous energy are lost in trying to fix labels on to the surface using transparent adhesive tape or self-adhesive labels.

In conclusion, the cage is the most important individual piece of equipment which has to be selected for the animal house, and time spent on careful consideration of all possibilities is well repaid.

C. GENERAL EQUIPMENT

Every animal room should be equipped with facilities for bottle cleaning and animal weighing. A number of satisfactory designs are available for routine bottle cleaning and no more need be said except that the reader is advised to talk to existing users of the apparatus selected, to ensure the equipment is satisfactory for his own specific purposes. Balances should be of the direct reading type with a scale covering the whole of the expected range in order to eliminate having to change scales during weighing. Well-damped units should be used to minimize fluctuations due to animal movement.

Other animal house equipment to be considered is as follows:

1. Feeding Utensils

Food should never be presented on the cage floor since this promotes contamination by excreta and a tendency to decomposition and is wasteful. Pelleted foods are conveniently fed in hoppers; these are either integral with the cage lid construction, baskets let into the cage top or large gravity hoppers affixed to the outside front of the cage with the opening passing into the interior (particularly with rabbit and guinea-pig cages or for feeding dry foods to cats and dogs). Wet mash and meal should be presented in hoppers or dishes contained in a larger outer dish to prevent soiling. Feeding dishes for dogs and cats should be heavy and robust; if the accommodation is suitable, presentation in hoppers above floor level reduces waste and labour; where groups of cats and dogs are maintained the number of food containers or length of hoppers should be adequate to prevent the more timid animals being denied their fair share. Where hay is fed this can be supplied in coarse-mesh racks.

The presentation of water still remains a problem to some extent. For almost all species the inverted bottle, with a spout integral with the lid, suffices for most purposes (Barber, 1970). The disadvantages lie in the need for frequent refilling and liability to flooding. Such bottles can be up to 500 ml in capacity and the spout made with an internal diameter of 6–9 mm and aperture 3 mm. The bottle may be made of glass or plastic and the most satisfactory spout is made of stainless steel; the latter is essential for use with rabbits or guinea-pigs. To prevent gnawing of the bottle a suitable metal shield should form part of the cage construction with a hole just large enough to allow passage of the spout.

A number of automatic watering systems have been tried with the common laboratory species but none is totally satisfactory. Among the more successful are poultry water fountains used for groups of cats or large guinea-pig floor pens; continuous-flow systems used in poultry batteries; automatic dispensers operated by a ball valve or lever actuated by the licking of the animal—used successfully with dogs and rabbits. The latter system, however, is subject to dripping caused by small particles lodging in the valve or by furring in hard-water districts.

2. Sterilizing and Washing Equipment

The need for an efficient cage-washing machine has already been discussed. Consideration should also be given to laundry facilities—particularly in self-contained units.

The nature and siting of sterilization equipment is a very important facet of animal-house design. Ideally the sterilizer should be open at both ends and separate "clean" from "dirty" areas. The concept of a sterilizing room with overlapping paths for clean and dirty equipment is obviously unsound and this has already been discussed in Section II.A. Where a single-ended sterilizer has to be used, precautions, such as double-bagging, should be employed to reduce the possibility of cross-contamination. For most purposes sterilization by autoclaving is the most satisfactory from the aspects of simplicity, speed and cost. Machines capable of operating at a maximum pressure of $2 \cdot 1$ kg/cm^2 are adequate and should be automatically controlled for cycles to include extraction of air, admission of steam, build-up of steam pressure, holding steam pressure, reduction of pressure and removal of steam and admission of filtered air. Completely automatic units save a great deal of labour and are usually provided with visual monitoring and warning systems related to failure to achieve pre-set limits and interlocking doors which permit only one door to be opened at a time and prevent the "clean" door from opening if the load has not been sterilized.

Ethylene oxide sterilization is used in some laboratories. This obviates the high temperature required for autoclaving but requires special consideration of wrapping materials and complete elimination of the fumigant at the end of the cycle. The process is also slower and potentially more hazardous to staff.

The complete sterilization of food represents a problem in that decomposition can readily occur with autoclaving and the physical condition of pelleted foods so altered that they become inedible. Pasteurization of compounded foods during the pelleting process suffices for all common organisms pathogenic to laboratory animals. For complete sterilization, when the food is to be introduced into germfree isolators, irradiation with $2 \cdot 5$ Mrad has proved successful although very costly. Doubt exists on the efficiency of a $2 \cdot 5$ Mrad level and some users insist on a $5 \cdot 0$ Mrad level. Many laboratories use diet sterilized in this way in barrier-maintained units, but the justification for this against the use of pasteurized diet is slight.

The equipment required for the sterilization of isolators is not extensive. The most commonly used method is to spray a 2 per cent solution of peracetic acid on to all internal surfaces, this sterilizes not only the surface but the contained air. All that is required is a hand-spray unit integral with the bottle containing the solution and pressure supplied by a hose from a nitrogen cylinder. Ethylene oxide and beta-propiolactone as an aerosol are also used.

Room sterilization, as distinct from the sterilization of room surfaces by disinfectants, can best be carried out by the use of a formaldelyde generator under saturated air conditions. The generators are cheap and simple to use and the saturated conditions are produced by a small electrically operated steam generating unit. This method produces complete sterilization, including the destruction of nematode eggs. As mentioned before, the construction of the room should enable it to be made airtight and electric fittings waterproofed before this method can be used.

3. Food Preparation

Where pelleted or complete commercially available foods are exclusively used, there is no requirement for food-preparation facilities. However, if a possibility exists for the large-scale administration of substances in the food, an area set aside for this purpose will be of value—particularly where the made-up diet is to be fed as a meal. An efficient mixer is required, which creates noise and vibration, and the inevitable dust arising from the meal rules out this operation in animal rooms. Many laboratories with a large population of cats or dogs mix their own diet for these species, using meat obtained direct from slaughter houses. These foods require mincing and cooking, and mixing with cereal meals or biscuit and mineral and vitamin supplements; such operations require the provision of a food-preparation area and suitable equipment. The area should be well ventilated. The choice of equipment is best made after consultation with units carrying out similar procedures; the equipment is expensive and maintenance costs heavy if it is not suited to the task—it is always cheaper to learn from the success or failures of others. If space permits, the installation of deep-freeze units permits bulk buying of frozen meats, so reducing cost and providing a buffer against fluctuations in supply; the minimum capacity should be 3 weeks' supply, but this may be reduced if the turnover is high and the contract for supply permits weekly delivery.

4. Furniture

Within the animal room furniture should be kept to the minimum. Apart from a sink, all that is required in a breeding room is a work top for weighing and recording and a drawer to hold the record book and cage labels. If post-mortem examinations are carried out within the room a second drawer is required for the dissection board and instruments. Cleaning equipment should be hung clear of the floor and a shelf provided for detergents and disinfectants. Floor-standing units should be avoided since these make cleaning difficult and provide inaccessible areas where dust and moisture may collect. In experimental rooms used for holding animals only, or where only occasional and simple experimental procedures are carried out, the same principles should be applied.

Where the experimental animal room becomes an extension of the laboratory and more complex apparatus has to be housed, with the consequent requirement for greater working space, movable tables will provide flexibility in the internal arrangement and enable the room to be easily converted for different purposes. When cupboard space is essential, units should be close-fitting and sealed to floors and walls; the internal fittings should enable cleaning to be carried out simply and efficiently. Wall cupboards are to be avoided, since the tops tend to get neglected during cleaning and may limit the flexibility of mobile rack arrangements.

5. Metabolism Cages

For metabolic studies in small animals a number of satisfactory units, constructed in glass, are available from laboratory equipment suppliers. Although these units are costly they have interchangeable parts, which reduces replacement costs due to breakage. Food and water are presented at the end of tunnels, which enables accurate consumption figures to be determined. While satisfactory for short-term experiments, long periods of confinement may stress the animal and affect experimental results, and may also contravene the requirements of the Home Office, in which event it is always advisable to consult the Inspector.

For larger animals, metal units are available commercially. Because of their bulk these units may cause storage problems and a collapsible type is to be preferred for this reason. The elimination of sharp corners and ease of excreta removal by using fibreglass for metabolism cage construction is a new development which may well prove worth while.

6. Clothing

The easy decision regarding clothing is to settle for standard white overalls or white jackets and trousers together with white caps. For male staff this is probably satisfactory and functionally serviceable where heavy work is involved. For female staff an excellent range of protective clothing is available in lightweight material made from synthetic fibres in a variety of colours. Although comparatively expensive this clothing is very durable, comfortable and contributes greatly to the morale of female staff. Where hosing down is necessary, waterproof nylon overtrousers and jackets are comfortable and serviceable. For the animal rooms, where the discipline is based on dry cleaning of rooms, lightweight plimsolls prove to be comfortable and cool.

7. Miscellaneous Equipment

Under this heading can be included such items as trolleys, which can be tailor-made for specific purposes at very reasonable cost by using slotted angle or box sections; plastic bins which may be of the refuse type or rectangular for easy distribution of food or bedding; high-pressure hose equipment; vacuum cleaning equipment, portable or centralized with individual room outlets; floor cleaners and polishers; electric fly killers; and killing boxes (see the UFAW handbook on humane killing). Again, the best method for deciding on such equipment is to shop around and see the items in use. In all cases, decisions must be made before completion of the brief to the architect, since the nature of these decisions may affect the details of construction.

IV Mechanical Aids

Labour costs represent a large proportion of the total expenditure in operating animal facilities and consideration should be given to mechanical means of reducing these costs. Although initial expenditure and maintenance may be high, when considered against the time saved, advantages will be apparent in many cases. The introduction of automation or mechanization can also produce a more satisfactory result by introducing the factors of thoroughness and elimination of human error.

Most of the mechanical aids used in animal houses have already been mentioned but will now be summarized for convenience.

1. Cage washers and sterilizers. These operations consist of a series of cycles which take considerable time to complete. To avoid the operator having to stand and watch when other duties could be undertaken this equipment should be automatic. The elimination of human error is extremely important in sterilization procedures and the minimum requirement should be for the continuous monitoring of temperature and pressure.
2. High-pressure hoses. These prove extremely useful where a number of the larger animals are maintained, whose accommodation requires frequent hosing down. The unit consists of a high-pressure lance connected to a mobile electrically operated pump, usually with a facility for mixing a detergent with the water. The water supply is connected by hose from either hot or cold outlets in the room, depending on the nature of the material to be removed. This equipment has proved satisfactory for dog kennels, where the faeces, possessing a high fat content, can prove extremely difficult and laborious to remove by conventional hosing.
3. Mobile water dispensers. In rooms where large numbers of water bottles have to be filled at frequent intervals—especially in small rodent rooms —much saving in time can be accomplished by the use of a water dispenser which can be brought to the vicinity of the cage. These take the form of a lever-operated tap, which can be hooked on a convenient point on the racking and connected by hose to a water supply. This ensures that the same bottle is always returned to each cage in addition to time-saving advantages.
4. Suction-cleaning apparatus. Where smooth-finish cages with solid floors are employed for small rodent breeding and maintenance, the use of vacuum cleaning is quick, thorough and simplifies the problem of waste disposal. The apparatus consists of an industrial vacuum cleaner which allows the expelled air to be filtered and solid matter to be collected in a plastic bag. The suction is applied by means of a flexible tube connected between the machine and a simple metal nozzle. For large animal rooms the apparatus can be moved on a trolley, but for smaller rooms the machine can be situated on the "dirty" side with a hose connection set into the intervening wall. In a building designed on the principle of a large number of small rooms, the number of machines required will depend on the minimum number of rooms requiring cleaning at the same time. These machines are very mobile and can be connected to external hose points set in the wall of whichever room requires their use. Where corridor space is restricted a centralized system can be

designed, the larger machines being able to accommodate four suction points at any one time (Foster, 1959). The suction pressure is such that cross-contamination is not possible. Anxiety over disturbance to the animals due to the noise level is often expressed; in practice, even when the machine itself is situated within the animal room, the nature of the noise has no detectable detrimental effects.

5. Floor-cleaning equipment. In view of the low cost outlay and long expected life of industrial floor-cleaning equipment, careful consideration should be given to such an investment if the size of the unit can provide a justification. Manual methods are time-consuming and since these are the least attractive of all animal-house tasks will become less liable to be overlooked and be undertaken more thoroughly if a mechanized cleaning technique is employed.

6. Bottle-cleaning apparatus. The techniques to be applied to maintain cleanliness in water bottles is a topic which can provoke heated argument. Those who achieve their goal by frequent rinsing and occasional sterilizing are probably in the majority over those advocating the use of bottle brushes and more frequent autoclaving. Whichever method is selected, equipment is commercially available which will scour the inside of the bottle with an electrically operated brush or provide high-pressure rinsing for the inside of the bottle. The value of such equipment must remain for the reader to decide.

V. The Atmospheric Environment

The atmospheric environment of the animal is vital to its health, comfort and ease of handling. This section will consider this aspect of the physical environment.

The permissible variability of temperature and humidity is known and accepted for all the common laboratory species; the achievement of these conditions can prove extremely difficult. Where groups of animals are allowed to run together within a large space—guinea-pigs in floor pens for instance—a satisfactory movement of air at the right rate, temperature and humidity in the vicinity of the animal can be achieved without great difficulty. Most laboratory animals, however, are maintained in cages and the provision of an adequate atmospheric environment at the locality of the animal requires very careful planning.

A. VENTILATION

The purpose of ventilation is to supply the animal with its oxygen and heat requirements, remove the products of respiration and the animal's own excess body heat, and maintain an acceptable level of humidity; if this is done efficiently the animal will be living constantly at its optimum temperature and humidity without draughts and without excessive odours from its body or excreta or variation in the normal chemical composition of air. In addition, adequate ventilation is required for the dilution of the bacterial and dust content of the air arising from the animal cages.

For low to moderate stocking densities (not more than an aggregate of 5 kg of rats per m² floor area) of rodents, about eight air changes per hour will normally be found sufficient, provided the air is well distributed. For higher densities (about 9 kg of rats per m² floor area), which the cost of building generally compels one to adopt, not less than fifteen air changes are necessary. The use of windows will not achieve this degree of ventilation, and therefore a ducted system should be regarded as the prime requisite in the design of new accommodation; the health of the animals is so dependent on efficient ventilation that it is difficult to see how expenditure on buildings can be justified without this cardinal feature.

Air is taken in from outside, filtered, its temperature and humidity adjusted, and it is then fed through ducts to the animal rooms, and to any other areas that require to be ventilated. To cause air to move along a duct it is necessary to create a pressure differential between the two ends of the duct, and this is provided by the fan that is moving the air. Similarly, if air is to pass through a room, at the rate of some fifteen room volumes an hour, there must be a differential pressure between the air in the room and the outside air, which will be nullified whenever the door is opened, or reduced if there are any cracks, gaps or holes in the wall. It is a matter of some importantce to decide whether the pressure in the room shall be higher than, equal to or lower than the outside. For breeding, and for the maintenance of healthy colonies, the room pressure should be higher, so that possibly infected air and dust are driven out rather than sucked into the room. But if the animals in the room possess pathogenic organisms, or it is important to contain animal odours, the room pressure should be lower than outside, so that air and particles do not escape. These differentials are achieved by adjusting the rate of flow in the input and outlet ducts. If a positive pressure in the rooms is being sought, an outlet duct can be dispensed with, in favour of a filtered grill or set of louvres, the resistance of which will maintain an internal positive pressure.

There are certain rules of thumb that are useful in calculating the volume of air that has to be provided by the ventilation system of a well-stocked rodent room. The animals on 1 m² of floor area will eat about 1 kg of food in 24 hours, and will require about 0·5 kg (0·35 m³) of oxygen for its metabolism. If the level of oxygen is not to fall, or the level of carbon dioxide to rise more than 1 per cent, the air in the room will need to be changed three times per hour.

Distribution of the air within the room is not less important than bringing the proper amount of air to it. Ventilating a room for human occupation is relatively simple, because breathing takes place almost entirely within a zone between 1·0 m and 1·7 m from floor-level. In such a room there are usually few furnishings at this level that will obstruct the free flow of air and, moreover, such rooms will not require such a high turnover of air as fifteen changes per hour. But a densely stocked animal room is quite another matter. All strata from 0·1 m to 1·8 m from the floor, and over the greater part of the floor area, need ventilating, in as uniform a manner as possible;

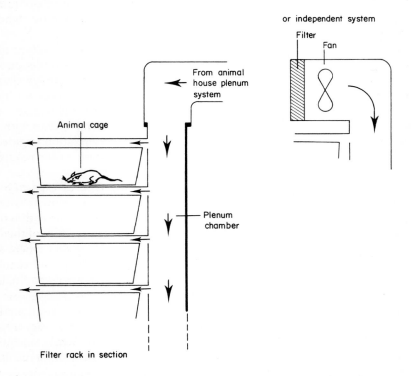

Fig. 6.10. Filter-rack system for ventilation of small animal cages.

and there will be shelves, racks and cages that will interfere grossly with the free flow of air within this zone. A study of animal rooms will indicate that, with conventional systems of ventilation, leaving air distribution within the rooms to the operation of secondary and tertiary air currents, even ventilation is almost never achieved, and sometimes not closely approached.

Adequate and even ventilation can be provided for small rodents by the use of the filter-rack system (Lane-Petter, W., 1970, a and c; Lane-Petter, M. E., 1970). This is based on passing filtered air at the correct temperature and humidity over the top of the cage from a plenum source built into the back of the cage rack (Fig. 6.10), which may itself be connected to the main animal-house ventilation system. This technique has advantages which could revolutionize the design and management of small rodent animal houses, if more extensive use in practice proves successful. The system is comparatively cheap, reduces the cost of building and operating a barrier system, prevents cross-infection between cages, reduces contamination of animals by organisms carried by staff, and provides ventilation at the point where it is required and therefore the overall demand is reduced. It could well provide the answer for a laboratory having a requirement for high-quality animals but possessing antiquated facilities where reconstruction is either impossible or too costly.

Whether or not a filter-rack system is used, but more especially if it is not, careful attention must be paid to the siting of input and outlet ducts. How they are placed, and what is the speed of air passing through them, may have a critical effect on the standard of ventilation in the room.

The plenum ventilation system should be flexible so that the input rate can be varied to individual rooms according to density and species within. The way this is achieved lies in the province of the ventilation engineer, and will depend on the capacity of the system in relation to the requirement.

The rate of air change overall will depend on the type of cage in use. Small rodents maintained in solid-floor cages possess a microclimate which may be quite different from the animal room itself; ventilation within the cage is due to a large extent to the eddies produced in the vicinity of the lid and the movements of the animals themselves. It is unlikely that where six to eight air changes per hour are found to be satisfactory, the rate of change within the cage is greater than two. Therefore, if the animals are maintained in open-mesh cages the rate of change can be reduced. The success of a proposed ventilation system can be envisaged by direct comparison with a similarly functioning operation, provided that animal density, cage type, rack arrangements and ventilation distribution are similar; where no comparisons can be made a good working estimate can be had from a distillation of the conditions existing in successful units. While small rodents show a capacity for adaptability, if a breeding or holding unit is to achieve long-term success, the ventilation conditions should be as near perfect as possible.

The ventilation requirements of other species depend mainly on bulk size, but other factors can enter into calculation of change rate. Although dogs require only about five changes per hour, the content of odour and excess humidity requires this rate to be increased to at least ten. Careful siting of ducts in kennel areas is required to ensure that the air movement does not by-pass the immediate environment of the dogs. The tendency of cats to succumb to respiratory infection can be minimized by a ventilation rate of ten to twenty air changes per hour; to reduce the effects of draught that this rate may produce, the cats should be provided with boxes or small hutches if a mesh cage is used.

B. TEMPERATURE

The temperature requirements of the various species to be housed can vary considerably. For small rodents the optimum temperature is 22–24°C; for poultry, guinea-pigs and rabbits 18–20°C; for cats and dogs not higher than 19°C; while marmosets require a temperature in the region of 27°C. Heat supplied by underfloor heating—electrical or piped hot water —or by radiators can produce continual problems in relation to ventilation, and fluctuations will occur. The most satisfactory method for the general animal house is through the forced ventilation system. Control is simplified by heating the air entering the duct system to the minimum requirement and providing additional heat in the ducts leading to individual rooms. This provides room variability which can be easily controlled by room thermostats. The essential requirement is the avoidance of sudden fluctuations and this can only be achieved, in conjunction with adequate ventilation, by the above method. The actual room temperature may be reduced below the animals' requirements where conditions are such that the microclimate is maintained at a higher level by the body heat of the animals. This is possible where solid-floor cages are employed and the animal provided with plenty of bedding material.

In a densely stocked room of rats and mice the heat of metabolism produced by the animals may raise the room temperature several degrees, even with an adequate number of air changes. If the ventilation fails, the temperature of the room may be expected to rise about 1°C per hour.

Where dogs and cats are provided with external runs and are therefore subject to considerable fluctuations in environmental temperature, a heat source in the kennel or hutch area is all that is required. The animal will orientate itself in relation to the heat source to satisfy its comfort. This heat source may take the form of infra-red lamps, tubular electric heaters (where space is confined) or underfloor heating.

The provision of heat is relatively inexpensive and to accomplish this a wide variety of systems may be used which can fit in with the method employed in the laboratory building as a whole. A requirement for cooling the animal room can be costly and this must be equated with an estimate of frequency of use together with an appreciation of the penalties incurred if it is not installed. Consideration should be given to the siting and method of construction of the animal rooms and positioning of air intakes for the ventilation system. The ventilation engineer should provide estimates, based on such data and annual temperature records, of maximum expected room temperatures and duration of periods when the optimum range is likely to be exceeded. This information can be used to assess the likely effects on animal breeding performance and health, and hence on the validity of experimental results and possible project delays. The effects of temperature increase above the optimum range depend on the absolute temperatures attained and the duration of the increase. In small rodents these are manifested by a reduction in mating frequency, then cessation of mating, abortion, poor lactation, litter neglect and abandonment, and reduced food intake and growth rate. Since these effects can be demonstrated at temperatures of 27–28°C, a level which may be maintained for some days in the British Isles, the resulting experimental delays or abandonment may be very costly. Over 30°C animal deaths may occur. Although it may only be required for a week or two in temperate climates the cost to research programmes must be weighed against the cost of installing a cooling system and must remain a local decision.

C. HUMIDITY

The control of humidity is necessary for the health of the animals, rats and mice in particular, and for providing good working conditions for the staff. A range of 50–60 per cent R.H. is generally acceptable throughout the animal house. The humidity should in any event not fall below 50 per cent R.H. and particular attention should be paid in rooms heated above the basic plenum level that this does not occur. These figures, of course, relate to the air in the room; the levels in the micro-climate in solid-floor cages will be higher.

In a well-stocked room of mice, there will be produced, on each m² of floor area, about 1·0 kg of water vapour every 24 hours. For rats, the figure may be between 1·5 and 2·0 kg per 24 hours. These figures assume that no condensation or absorption takes place, an assumption that, by reason of inadequate ventilation, is often not true. If this amount of water vapour is to be removed at such a rate that the relative humidity of the outgoing air is not more than 10 per cent higher than that of the incoming air, something between ten and twenty air changes per hour will be necessary. If hosing down of floors is frequently practised, the level of humidity will rise much higher than 10 per cent and only slowly subside. Thought should also be given to the effect on the outlet ducting of this high level of humidity.

D. FILTRATION

Two aspects of air filtration should be distinguished and considered on their own merits. The removal of dust particles from the input air in a forced ventilation system is essential. Bacteria can be carried on such particles and they should therefore be eliminated; also, the presence of dust can, after a period of time, clog ducts and louvres and cause wear in the mechanical part of the system; and it adds to the difficulty of room cleaning. The removal of dust particles down to a size of about 4μm in diameter, which is the lower limit of those liable to be carrying bacteria, is a comparatively simple procedure requiring a large area macrofilter which has a low resistance to the passage of air and therefore has little effect on the fan load. Such filters are cheap and easy to replace. The frequency of replacement depends on the situation of the animal house; filters in an urban area will obviously require more frequent attention than in a rural environment.

A 4 μm filter will not remove bacterial or fungal spores, nor viruses that are not attached to dust particles, although a number of such smaller particles will be trapped in such a filter, and those that get through may not be important. Absolute filtration requires the use of much finer filters, removing all particles down to a diameter of 0·1–0·3 μm. Such microfilters have a high resistance, requiring a higher rating of the ventilation unit in order to maintain the necessary level of throughput. Although such filters are placed behind a macrofilter (i.e. nearer to the fan), they tend to clog more easily and can serious affect the pressure balance in animal rooms if not changed frequently. This can be overcome by automatic devices which are operated by the pressure differential on either side of the filter and are based on a roll of filter material which moves along as it becomes clogged. These devices are expensive, as are the filters themselves, and the value of microfiltration must be weighed against the expected advantages. The employment of a recirculation system to economize on temperature control requires the process of microfiltration and deodorization as essentials, and the cost saving becomes so small that this technique is hardly worth the problems that may be encountered and additional maintenance required. Where an animal house is free-standing and not in a heavily built-up area the air dilution should be such that microfiltration of organisms likely to be pathogenic to laboratory animals should not be necessary. The arguments relating to apparent and inapparent infection discussed in Chapter 5 are particularly relevant to the consideration of microfiltration. Where other animal houses with separate ventilation systems are in close proximity, particularly where a high standard of husbandry is not employed, or in heavily built-up areas, a justification might be made for microfiltration.

In conclusion, if proper attention is given to the provision of a well-designed filtered ventilation system, giving adequate and flexible control of temperature and humidity, the rewards will be great in terms of animal health and agreeable environment for staff. It need hardly be said that the consideration of such a system should be made at the start of animal-house planning and not left as an afterthought, since this facility is integral with the construction of the building.

E. LIGHT

The inclusion of windows in animal rooms is associated with the preference of staff in providing an amenable working environment. However, certain disadvantages do exist. Direct sunlight must not be allowed to fall on to cages; temperature regulation may be made difficult; some species require a closely controlled photoperiod for a satisfactory breeding performance. Apart from the latter, the objections can be overcome by the use of double glazing and external or internal shades or blinds. The improvement to the working environment by the inclusion of windows is much greater in small rooms than in large, since in the former only one or two technicians may be working and greater attention must be paid to their morale. Provided other considerations of lighting are adequate, from the animals' point of view the exclusion of windows is no disadvantage. The Home Office, however, do recommend that equidae, cats, dogs and primates maintained on long-term experiments should be given the facility of exposure to natural light.

The provision of light for breeding, growing and experimental animals, together with means for controlling the photoperiod, are essentials for all animal rooms. There would appear to be variations in the recommendations of many authors regarding intensity and duration. The design of animal houses and arrangements for the support and distribution of cages depend mainly on the important considerations of ventilation, disease control and husbandry; the result of implementing decisions based on these criteria makes uniform lighting for all animals an extremely difficult task. Where sacrifices have to be made they should fall on the least vital consideration, in which category, in the absence of contradictory data, the concept of even lighting falls. Porter (1968) relates a case where light intensity varied from 20 lumens per ft^2 on benches below a light fitting to $2 \cdot 5$ lumens per ft^2 on the lowest animal cage shelf in a corner furthest removed from the fitting. The light within the cages would be of a much lower intensity. Excess intensity should be avoided (Porter, Lane-Petter and Horne, 1963, b) and, with albino animals, shading may have to be provided for cages at the top of tall racks. The aim when considering intensity should be to provide adequate light for good staff working conditions without the introduction of glare and 20 lumens per ft^2 would appear to be a generally accepted level. The provision of artificial light by the use of fluorescent fittings results in shadowless and cheerful conditions for the staff, provided the siting of such units results in a reasonably uniform intensity.

The photoperiod should be controlled to provide uniform light and dark periods throughout the year. The inclusion of windows, admitting natural light to animal rooms, will make this impossible unless the uniform light period required is equivalent to or greater than the maximum period of daylight. However, the provision of light during winter is of far greater importance than the exclusion of light during summer and for most purposes extended periods of daylight can be ignored. A lighting cycle, operated by time switches, of 12 hours daylight and 12 hours darkness is satisfactory for most species. It is important that the light cycle in experimental rooms is the same as that in the breeding unit from where the animals originated, particularly where reproductive studies are undertaken. Sophisticated systems for light control have been suggested whereby switching on and off is performed gradually through a "twilight" period, or a dual lighting system installed for optimum levels of intensity for working staff and animals (Cobb, 1963); since little is known of the optimum conditions for most laboratory animals and the achievement of uniform intensity to all cages in a room would result in very low animal densities, these suggestions are of little practical significance.

In the absence of valid experimental data relating to animal preferences the selection of lighting colour is largely based on personal preference. The advantages of fluorescent over filament lighting have already been mentioned and with this method the warm white tube provides satisfactory working illumination without glare.

VI. Pests

It should not be possible for any pests to enter the animal house, since they may be vectors of disease and present a real hazard to the health of laboratory animals. In addition, in accommodation for animals such as dogs, not kept under rigid barrier conditions, the presence of wild rodents may result in considerable structural damage to false roofs, insulation and even electric wiring and conduits.

Rodents can be excluded by making external walls smooth to a height of 60 cm from the ground, and rodent barriers installed in doorways and entrance corridors (Fig. 6.11). Where entrance by trolleys is required doors should be metal clad to a height of 60 cm and a removable metal barrier 60 cm high installed immediately behind the door, fitting into side and floor slots. Doors to animal rooms should also be metal clad to prevent escaping laboratory animals from passing into other rooms. External runs should be provided with solid smooth outer walls of at least 60 cm in height with a rodent barrier above. Drains and gullies must be rodent proof, together with entrances to duct systems.

Insect pests provide different problems. In the initial fumigation of a new animal house, there may be hidden corners, under shelves or in pipe sleeving, where insect eggs or pupae are present, which are not accessible to the fumigant, at least in sufficient concentration to kill them. In consequence, although all exposed surfaces are completely sterilized by such fumigation, flies will hatch out later, to general consternation. Care should therefore be taken to search for such hiding places and deal with them, preferably before fumigation.

Infestation with cockroaches, bed bugs and crickets occurs in older animal houses, and has been seen even in comparatively new ones. This kind of infestation can only occur where there are cracks or other places of concealment for the insects, and in the construction of a new animal

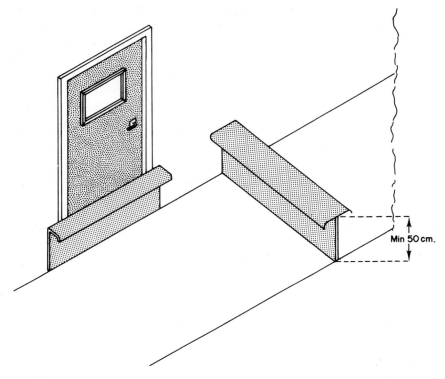

Fig. 6.11. Construction of metal rodent barriers shown in cutaway view of corridor and external door.

house such places should not be permitted. Perhaps the most satisfactory way of dealing with an existing infestation, and of preventing it occurring or recurring, is to apply an insecticidal lacquer to surfaces where the insects may crawl or alight. These lacquers, when walked over by insects, release microcrystals of insecticide, which the insects carry to their lairs, and the infestation can thereby be rapidly and completely eliminated. Since the lacquer does not come in contact with the laboratory animals, it presents no danger to them.

Flying insects can be controlled by the use of thermal aerosol insecticide generators, which produce a continuous or intermittent aerosol of insecticide in the room. The use of such generators in an animal room is not likely to present toxic doses to the animals, but they will absorb some insecticide and thus be rendered unsuitable for many kinds of experimental work. Aerosol generators are therefore not recommended in animal rooms, or in food or bedding stores.

Flying insects may be excluded by the use of double-entry doors, fly screening and electric devices placed in corridors which attract insects and kill them by electric shock. Fly screening is particularly important in the control of disease in cats; while adult cats will leave flies alone, kittens cannot resist chasing and even eating these pernicious disease vectors.

Food and bedding stores must be subjected to the same preventive measures as the animal rooms. It is particularly important to ensure that, in these stores, the material is used in rotation, so that a stagnant corner of food or bedding does not accumulate at the back of the shelves. Such accumulation would be an open invitation to all manner of pests; not to mention a welcome living area for any escaped mice or rats from the animal house.

NOTE ADDED IN PROOF

Insecticidal Lacquers

A useful leaflet on the use of insecticidal lacquers has been produced by the Ministry of Agriculture, Fisheries and Food, Pest Infestation Control Laboratory, Hook Rise, Tolworth, Surbiton, Surrey. It is entitled *The Control of Cockroaches and Pharaoh's Ant Using Insecticidal Lacquer*, and it includes a list of suppliers of these products.

CHAPTER 7

Nutrition

I. Introduction

In the conditions of a laboratory, animals are entirely dependent for all their nutritional requirements on what is offered them. They have no opportunity of supplementing an incomplete diet by foraging, as has the wild animal, or the animal not kept in such strict and hygienic captivity. This total dependence is equally true in cases where refection (rats) or coprophagy (rabbits) are established habits of the species.

Species vary in their nutritional requirements. The needs of rats and mice are close enough, qualitatively, to make it possible for the same diet to serve for both, but this is exceptional. The general rule is that each species has its own feeding regime.

The dietary requirements are not the same at every period of the animal's life. Rapid growth, pregnancy, lactation, and even an environmental temperature below the optimum, will all make demands on nutrition above the basic level required for bare maintenance. So also may the presence of certain infections or infestations.

Food preferences are exhibited by all animals, as well as a distaste for certain foodstuffs. These preferences should be regarded as a useful but by no means infallible guide to the needs of the animal. Preference for or toleration of a particular diet may be largely the result of habituation; and a change to a new diet, even a better one nutritionally, may be resisted for a period by the animal.

The theoretical composition of a diet, as worked out from tables of average values for the various ingredients, may differ substantially from the actual composition of the food as eaten by the animal. This difference arises from the variation in composition of individual batches of ingredients, from the effects of storage, from interaction between ingredients after mixing in a compound diet, and from other causes. But it is what the animal eats that counts, not what it was intended to offer it.

Disturbances of nutritional origin may be due to a deficiency in overall quantity, or result from shortage, absence or excess of particular ingredients. Too little food will lead to poor growth of young, loss of weight in adults, disturbances of reproduction, and often enough a lowered resistance to disease. A moderate restriction of total food intake may, however, have no adverse effect on reproduction, and may even prolong the breeding life of the animal. It may also enhance resistance to certain infections, and prolong the life span of healthy animals.

Gross imbalance between the bulk ingredients—protein, carbohydrate, fat, indigestible residue—and deficiencies or excesses of the accessory factors—vitamins, minerals—are the things that constitute bad feeding. But the signs of bad feeding are not necessarily the death or serious illness of the animals, nor are they always quick to appear. For example, a moderate deficiency of vitamin E may not show itself for two or three generations and then only in reduced fertility. An excess of carbohydrate, with a relative shortage of protein, may produce fat, healthy-looking animals, but the imbalance may once again show itself in reduced fertility.

This points the necessity of having some means of assessing a diet. The simplest way is to measure weight gain during a period of normally rapid growth: say from weaning to sexual maturity. But this gives remarkably little information. It overlooks the possible effects on fertility, the long-term anatomical and physiological effects, and the possibility that a heavier animal may be heavier only because it is abnormally fat or dropsical.

Observation of the complete generation life-cycle is more useful; that is, from the weaning of one generation of animals to the weaning of their progeny. Sometimes it may be necessary to carry the observations through two, three or four generations.

Reproduction, especially in a prolific animal like a mouse or a rat, makes very great nutritional demands. A diet that will support health, normal growth and even a moderate rate of reproduction may yet prove inadequate when reproduction is intensified. Hence the need to test a diet under the most exacting conditions that are likely to be encountered, including a spell of intensive reproduction.

There are differences in dietary requirements, not only between species, but between different strains of the same species and within the same strain in different environments. It is therefore desirable, in assessing a diet, to feed it to more than one strain and, if possible, to test it simultaneously in two different laboratories.

II. Dietary Requirements

All animals need certain basic substances in their diet, but in different proportions. The basic requirements are protein, fat, carbohydrate, vitamins, salts and minerals. Higher levels of food intake are required during growth and lactation, and a higher proportion of protein in the diet is also required at these times. For some species, a minimum level of fibre is also necessary.

Certain constituents of the diet may affect the requirement for other constituents. For example, an increase in the selenium content may have a sparing effect on the vitamin E requirement, and the relative levels of calcium and phosphorus have to be within certain limits; the same is true of the ratio of methionine to cystine, and also of phenylalanine to tyrosine.

There are several published recommendations of nutritional needs of laboratory animals. Among the more useful are Cuthbertson (1957); Publication 990 of the US National Academy of Science—National Research Council (NRC, 1962); *Laboratory Animal Handbook 2* (LASA, 1969); Drepper and Weik (1970) and Weik and Drepper (1970); the section on nutrition in *Husbandry of Laboratory Animals*, and also *Nutrition and Disease in Experimental Animals* (see Bibliography). Joubert (1967) has summarized information from a number of sources.

Table 7.1

Recommended levels of amino-acids in the diet of rats and mice (LASA, 1969), and of mice, rats, hamsters, guinea-pigs, dogs cats and monkeys (Weik and Drepper, 1970), expressed as a percentage of total protein

	LASA	Weik and Drepper
Arginine	5·0	6·0
Histidine	2·5	3·0
Lysine	6·0	9·0
Tyrosine	4·0	10·0
Phenylalanine	5·0	10·0
Tryptophan	1·5	1·2
Methionine	4·5	6·0
Cystine	4·5	6·0
Threonine	4·0	4·5
Leucine	8·0	7·5
Isoleucine	5·0	6·0
Valine	5·5	5·5

A. PROTEIN

Protein is an essential constituent of any animal diet. According to Weik and Drepper (1970) the percentage of crude protein in the diet can vary from 7 per cent in a maintenance diet for mice, rats and hamsters, to 45 per cent in a breeding diet for cats. Most laboratory animal diets contain 18–25 per cent of crude protein, except diets for cats, in which much higher levels are necessary.

The protein can be of vegetable or of animal origin, or a mixture of both. It is important, however, that it contains all the essential amino-acids—that is, those that the animal cannot synthesize itself—in the right proportions. For most laboratory animals there are about twelve essential amino-acids.

Table 7.1 shows levels of these twelve amino-acids, recommended in *Laboratory Animal Handbook 2* (LASA, 1969) and in Weik and Drepper (1970).

There is a very close measure of agreement between these two sets of recommendations. The biggest discrepancy is in the recommended levels of lysine. This is one of the least stable of the amino-acids, which may account for the higher level proposed by Weik and Drepper.

The main sources of dietary proteins are casein (from skimmed milk powder), soya beans, fish meal, meat meal and, for some special diets, eggs. There are lesser amounts of protein in cereals and lucerne.

For a full discussion of the relationship between protein and calorie requirements, see Payne (1967).

B. FAT

Fat is a necessary constituent of diet, in order to provide energy and as a vehicle for fat-soluble vitamins and essential fatty acids.

Recommended fat levels vary from 2 to 8 per cent of the diet, except in diets for cats, where levels of 10–30 per cent are necessary. Much of the fat can be of animal origin—tallow, for example—but a proportion must be of vegetable origin to ensure adequate levels of essential fatty acids.

C. CARBOHYDRATE

The cheapest constituent of any animal diet is, weight for weight, carbohydrate. This is a source of energy, and in all diets (except those for cats) forms more than 50 per cent of the total diet.

Sources of carbohydrate are mainly cereals. Some carbohydrate may be added in the form of sugar or potato starch.

D. VITAMINS

All animals have requirements for a number of vitamins, and recommended levels are given by LASA (1969) for rats and mice: by Weik and Drepper (1970) for mice, hamsters, guinea-pigs, rats, rabbits and monkeys; and for dogs and cats: and by Jelínek (1967).

Table 7.2 gives a comparison of various recommendations, in respect of mice and rats.

The recommended levels of several vitamins for dogs and for cats are considerably higher than those shown in Table 7.2. It is, however, possible to give breeding cats a depot injection of vitamin A in a vehicle from which the vitamin is slowly released.

For guinea-pigs and for monkeys, vitamin C (ascorbic acid) is also needed, at the rate of 1 g per kg diet, or 250 mg per l of drinking water.

Some vitamins are liable to undergo chemical destruction in the made-up diet, on storage or in the process of sterilization. The most labile is vitamin B_1 (thiamine), and autoclaving or heavy γ-irradiation may destroy 75 per cent of it. Ascorbic acid is also liable to undergo partial or complete destruction on keeping. Vitamin A, and also vitamin E, are subject to oxidation, especially in the presence of rancid fats. Most of the other vitamins are reasonably stable, and will sufficiently withstand sterilizing treatments without serious destruction.

Table 7.2.

Recommended levels of vitamins per kg of diet for mice and rats (LASA, 1969; Weik and Drepper, 1970; Jelínek, 1967)

	Unit	LASA	Weik and Drepper	Jelínek rats	mice
Vitamin A	i.u.	5000	5000–15,000	400	300
Vitamin D$_3$ (cholecalciferol)	i.u.	300	300–500	not required	
Vitamin E	mg	60	60–150	30	20
Vitamin K$_3$ (menaphthone)	mg	1·5	2–10	not required	
Vitamin B$_1$ (thiamine)	mg	4·0	4–10	1·25	20
Vitamin B$_2$ (riboflavine)	mg	5·0	5–10	2·5	8
Vitamin B$_3$ (pantothenic acid)	mg	12·0[a]	12–50	10	6
Vitamin B$_6$ (pyridoxine)	mg	6·0	6–15	1	2
Nicotinic acid	mg	10·0	15–50	not required	
Biotin (vitamin H)	μg	0	200	required	?
Folic acid	mg	0	5	required	10
Vitamin B$_{12}$ (cyanocobalamin)	μg	5·0	5–30	required	1–2
Choline	g	1·0[b]	1·0	1	1
Essential fatty acids	g	not given		10	10

[a] as Ca pantothenate.   [b] as choline chloride.

E. SALTS AND MINERALS

LASA (1969), Weik and Drepper (1970) and Quarterman (1967) make recommendations for levels of salts and minerals, which are given for comparison in Table 7.3.

Although most of these minerals will be present in sufficient amounts in any suitable mixed diet, this cannot be relied upon, especially for iron, sodium chloride and calcium. It is, therefore, better to add a mineral supplement to the diet. This is conveniently done in the form of a vitamin and mineral premix.

F. FIBRE

In all diets made from natural ingredients there is a certain level of fibre, consisting in the main of cellulose from plant cell walls. The fibre levels in the diets of ruminants, and of rabbits and guinea-pigs, should be relatively high—Weik and Drepper (1970) recommend 8–18 per cent—because these animals are able to digest cellulose in the gut; indeed, it is a

Table 7.3

Recommended levels of salts and minerals per kg diet for rats and mice (LASA, 1969), for mice, hamsters, guinea-pigs, rats, rabbits and monkeys (Weik and Drepper, 1970) and for rats (Quarterman, 1967)

		Unit	LASA	Weik and Drepper	Quarterman
Boron	(B)	mg	0·04	a	a
Calcium	(Ca)	g	6	a	a
Chlorine	(Cl)	mg	500	a	a
Copper	(Cu)	mg	5	5	5
Fluorine	(F)	mg	0·04	0·5	0·007
Iodine	(I)	mg	0·02	0·5	0·15
Iron	(Fe)	mg	50	100	25
Magnesium	(Mg)	mg	500	a	a
Manganese	(Mn)	mg	50	50	50
Molybdenum	(Mo)	mg	0·04	a	0·02
Phosphorus	(P)	g	5	a	a
Potassium	(K)	g	5	a	a
Selenium	(Se)	mg	0·04	a	a
Sodium	(Na)	g	5	a	a
Sulphur	(S)	mg	300	a	a
Zinc	(Zn)	mg	12	20	12

[a] not given.

necessary part of their diet. For mice, rats, hamsters, dogs and monkeys, fibre levels of 4 or 5 per cent should not be exceeded. Rats are capable of breaking down some cellulose in their gut, by bacterial action, and these bacteria may at the same time synthesize some of the B-vitamins. The rat therefore refects—that is, it eats some of its own droppings, and thus extracts the B-vitamins from them. Coprophagy in rabbits probably serves a similar function, and in them the high fibre content is essential. Only cats are intolerant of fibre levels above 2 per cent.

Cellulose from the cell walls of cereal grains is soft, and if not digested is excreted unchanged, without any damaging effect on the gut. But some types of plant fibre can do serious damage, especially to the guts of young rats and mice. Sawdust particles, in particular, which contain the indigestible lignin as well as cellulose, are spiky and, if ingested by young rats or mice, may penetrate the mucosa lining the gut and either perforate it or lead to a cellular infection. In either case the animal is likely to die, from peritonitis or from obstruction. The hard husks from barley, and even from oats, if not finely ground, can also have a similar effect.

III. Composition and Presentation of Diets

The dietary requirements discussed in the previous section have to be translated into food suitable for the animals to eat. All laboratory animal food
must be of the right composition to supply all nutritional needs; its physical form or presentation must be such that the animals can eat it; it must
meet whatever demands are likely to be made on it for transportation and
storage; and it must be palatable. These are invariable requirements;
to them may be added the frequent necessity to sterilize or pasteurize the
diet without destroying its completeness or in other ways rendering it
unfit.

When all these requirements have been met, there still remains the problem of making the diet up with reasonable economy, from ingredients that
are not only readily available at all seasons, year after year, but which are
themselves not liable to serious variations in quality or nutritional value.

The compounding of laboratory animal diets is not so simple as the
preparation of farm animal feeds. Most diets are compressed into pellets,
but a few are in expanded form. Many of the laboratory animal feed
formulae do not easily lend themselves to pelleting, while the mixtures
that pellet well may not be the best formulae. Unquestionably there is need
for a great deal of technical skill and knowledge in the business of making
satisfactory laboratory animal feed pellets; not every feed manufacturer
is able to handle it, although some try, and fail.

A. COMPOSITION

Laboratory animal diets are made up, for the most part, from natural
ingredients that have undergone some processing. The bulk is cereal, which
may include wheat, oats, barley and maize.

All of these may vary in their protein content, and in the proportion
of fibre, within appreciable limits. Maize is perhaps the least variable. The
grains may be as threshed, or with more or less of the husk, the bran or the
germ removed. When milled, they may be coarse or fine ground.

The grain, or even the meal, may have residues of insecticides, fumigants
or other chemical contaminants. It may also contain evidence of wild rodent
infestation, in the form of hairs, powdered or dried droppings, and any of
the infections that might accompany such contamination. Occasionally,
too, cereals are affected by fungi, among which ergot (*Claviceps purpurea*)
and aflatoxin (from *Aspergillus flavus*) are not unknown.

In compounding a feed, therefore, considerable care needs to be taken
that the cereals chosen are clean, of a specified type or quality, and can be
obtained regularly. To compensate for fluctuations in the quality of successive batches or crops of a cereal, some compounders mix more than one

cereal together, in the hope that a deficit in a batch of one will be compensated by the others. But this is less sure than paying attention to the quality of the batches of the chosen cereals, of which batch analysis is to be preferred to leaving it to the luck of a mixture.

The protein level in diets is raised by the addition of fish meal, meat and bone meal, dried milk (usually skimmed) or casein, and soya-bean meal; any or all of these. All these protein sources are excellent, but all have certain disadvantages, which should be well known to the compounder.

Fish meal containing a high proportion of herring or other oily fish is likely to lead to oxidation of the vitamin A or E content of the mixed diet. White-fish meal is therefore to be preferred; alternatively, the addition of an antioxidant to the diet is necessary. The process of drying and milling the fish can denature some of the protein, which will make it less useful as a food.

Meat and bone meal, sometimes mixed with blood meal, provide first-class protein and a reasonable level of calcium. They may also contain dangerous bacteria, including salmonella. Processing can also lead to some denaturation of the protein.

Milk powder, or unrefined casein, is rather more expensive than either fish or meat and bone meal, but is likely to vary less in quality. It is, however, subject to some alteration on storage, and can become seriously denatured.

Soya-bean meal contains first-class protein; it also contains an enzyme, trypsinase, that will inactivate the animal's digestive proteolytic enzyme, trypsin. The trypsinase is normally inactivated by heating the soya-bean meal. Soya-bean protein is also reported to antagonize vitamin K, but this is not likely to be a serious contraindication for its inclusion in a diet.

Lesser sources of dietary protein are from the cereal content of the diet, and from yeast. All these sources of protein may yet produce a diet that is low in the sulphur-containing amino-acids, methionine and cystine. It may therefore be necessary to supplement the protein with added methionine.

The cereals and the fish, meat and bone meals all contain some fat. The milk powders and the soya-bean meal are likely to contain little or no fat. Extra fat may be added in the form of corn oil or other vegetable oil; or of tallow, which is a mixture of animal fats, chiefly beef and mutton. There must be a minimum level of unsaturated fatty acids: the LASA (1969) recommendation is that not less than "1 per cent of the dietary calories should be supplied as linoleate". The fat in fish meal, or the vegetable oil, will easily ensure this.

More important, perhaps, is the avoidance of rancid oil or fat, which can cause serious digestive disturbances to the animals that eat it. The compounder should specify, and verify, iodine levels in all the fat added to the diets.

Vitamins and minerals are most conveniently added as a premix.

B. PRESENTATION

Before the advent of compressed pelleted diets, compound feeds were presented to the animal either as dry meals, or as mashes consisting of meal with water added to make a stiff paste. Dry meals are wasteful but they can be used if fed from suitable hoppers. They have the advantage that experimental mixtures can be easily made up, and that standard diets in meal form can be medicated or modified by the addition of other substances for experimental purposes.

Such meals may be mixed with water to form a stiff paste or mash, and fed in this form. Mashes sour quickly, especially at animal-room temperatures, and within a day or two will also grow mould. Excess mash tends to get lost in the bedding, or fall through the grids of wire-bottomed cages. They thus need to be fed daily, 7 days a week. Wet mashes have little to recommend them.

A wet mash may, however, be dried in a low-temperature oven, to form a hard cake or biscuit, the water content of which may be as low as 5 per cent. Such a dried mash is useful in preparing small batches of experimental diets. A simple method is to roll out a stiff paste in a shallow cake tin (1 cm deep: commonly called a "Swiss roll tin") and dry in the oven. If the paste is scored as it dries, it breaks up when dry into convenient cakes or biscuits. The keeping quality of such a cake is good.

Bennett (1969) has described a method of making pelleted diets in experimental quantities, using a kitchen mincing machine to extrude a cylinder of paste, which is then dried in warm air.

However, for the great majority of purposes, laboratory animals are fed on compound compressed pellets. These are made by mixing the ingredients together, sometimes injecting steam or hot water, and forcing the mixture through a die. The mixture is heavily compressed in the die, in which a lot of heat is generated—enough to kill most or all of the vegetative bacteria and parasite ova present—and emerges in a rod of cylindrical or other shape. The rod is broken off to form fairly uniform pellets.

The process of pelleting calls for a certain degree of technical skill. The meal is made to stick together into pellets because it contains binding agents such as molasses, casein or a number of other substances that have been specially developed for this purpose. Difficult mixtures to bind into pellets are those with a high fat or fibre content, coarsely ground cereals or low protein.

Whether the pellet turns out to be hard and well formed, or soft and friable, depends not only on the composition but also on the efficiency of the pelleting machine. These machines are of various patterns, and they are costly. Some make much better pellets than others. Nearly all can only handle batches of mix of one or more tons, although there are a few available that will handle smaller quantities.

Another method of making pellets depends on mixing the ingredients under pressure, with the injection of steam, which will almost if not completely sterilize the mixture. It is then extruded into the atmosphere, where the release of pressure causes the steam in each pellet to expand and create a spongy mixture. Pellets made in this way may then be polished by rolling them in a drum. Fat, the level of which must not be high in the mixing chamber, may be sprayed on after extrusion and expansion: since the texture is spongy the fat is readily absorbed.

Expanded pellets are resistant to crumbling, and so store and transport well. They have a specific gravity of little more than half that of compressed pellets, and so take up nearly twice the space in the store room and in the food hopper.

For some purposes canned feeds are available, especially for dogs and cats. It is the most expensive method of presentation, and is almost confined to meat and fish products.

From the animals' point of view pelleted diets have very much to recommend them. They are usually placed in wire hoppers in the cage. and the animals gnaw them through the wires. They do not, therefore, become soiled with bedding and droppings, as they would do if placed loose in the cage or in an open dish.

There are, however, one or two points to watch. Badly made pellets are likely to be friable, leading to much wasted powder in the bottom of the bags. Such friable pellets are mealy to the taste, and the animals will also waste a substantial proportion of them in the cage. On the other hand, very hard pellets may be too much for young animals beginning to eat solid food; they may have unreasonable difficulties in gnawing them through the wires of the hopper.

Since pelleting machines are expensive few laboratories or breeding units can afford to instal their own. They are dependent on the good faith of the compounder. Unless the compounder not only knows his pelleting technology but also has more than a passing acquaintance with the principles of nutrition, his product is likely to be neither uniform nor complete. The appearance of the pellet gives no clue about what went into it.

Many compounders are reluctant to put on their bags of feed the date of manufacture. Since all these diets deteriorate on keeping, this would seem to be an important item of information for the user. A compounder who refuses to date his bags invites grave suspicion of his methods and reliability.

Pelleted feeds are almost always offered to the animals in hoppers or wire baskets. Hoppers deliver the pellets by gravity as the animal consumes them. With wire baskets the animal chews the pellets through the wire. It is important to ensure that the spacing of the wires is correct, in relation both to the size of the pellets and also the size of the animal's muzzle. It is also necessary to make sure that even the smaller animals in the cage can reach the food basket; and that the food cannot be fouled by the animals, because the basket is wrongly placed.

C. PALATABILITY

Since laboratory animals are totally dependent for all their nutritional requirements on the food that is offered them, it might be thought that palatability is of little or no importantce. There is no evidence to prove whether one diet is preferred to another by reason of its greater palatability, or even if laboratory animals like eating, in the sense that a gourmet likes eating. But in most breeding colonies the intensity of breeding is high, and therefore the food intake must also be high. A diet that is readily eaten—whether it is liked is therefore irrelevant—is therefore likely to produce better growth and reproduction than a diet that appears to be unpalatable (Lane-Petter, W., 1970, b).

Generally speaking, compressed pellets seem to have a moderate to low level of palatability, and this can be confirmed for human taste by eating them. It is not so much the taste of fish as the mealiness which is objectionable. Guinea-pigs eating pellets will often pass from food hopper to water bottle repeatedly, as if the pellets needed a lot of washing down.

Dried paste, or wet mash freshly prepared, seems to be more palatable, but feeding trials in mice have failed to demonstrate that, by the standards of growth and reproduction, the more palatable diet gives better results. Expanded pellets are eaten readily, and appear to be at least as palatable as compressed pellets, if not more so.

Occasionally a diet will be rendered very unpalatable because it has become contaminated with some offensive substance. In one such case, a batch of compressed pellets that had come from a new machine was totally refused by a colony of rats, and the cause was considered to be the residue of oil in the machine, which gave the pellets an unpleasant taint. Surprisingly, however, a colony of rats of a different strain ate the same diet with no difficulty.

IV. Assessment of Diets

An animal must eat food in order to provide energy for moving and for keeping warm; in order to grow; and in order to reproduce and raise young. An active animal will use up more food calories than a sluggish one; and so will an animal in a cold environment rather than a warm one. Some animals have more efficient digestive systems than others, and can extract more of the useful contents of the diet. Disease, even if clinically silent, can impair an animal's digestive efficiency, or interfere with the absorption of nutriments. There can be considerable differences in the digestive efficiency of different animals of the same species or strain; and such differences exist in the same animal at different periods of its life.

A diet may be regarded as optimal if the animal eating it grows and reproduces at its maximal potential level. This level may not be known, and in any event will vary in different environmental circumstances, so that no absolute assessment of a diet is possible. It can only be compared with other diets, and be either better or worse than them, or equally good or bad.

A diet is unacceptable if the animals eating it develop signs of nutritional deficiency, such as hypovitaminosis or other deficiency diseases, or if it fails altogether, or nearly so, to support a reasonable rate of growth and reproduction. Some deficiency diseases, such as avitaminosis E or K in rats, may take weeks or months to develop, and may even take two or three generations of breeding to become evident. Others, like avitaminosis C in guinea-pigs, will show up rather quickly, usually because the scorbutic guinea-pig is particularly susceptible to bacterial infections and will sicken and die from any that are present.

Protein deficiency, and in particular specific amino-acid deficiencies, will be reflected in poor growth and low resistance to infection, and sometimes by specific signs such as depigmentation, falling hair or skin changes. Excess of saturated fats, or deficiency of unsaturated fatty acids, may be associated with atherosclerosis in rats. For a discussion of trace-element deficiency, reference should be made to Quarterman (1969).

When devising a diet there must be a practical method of assessing it. Clearly, a diet that causes deficiency diseases is not satisfactory. But a diet may still be poor, even in the absence of all recognizable signs of such deficiencies. The animal may find it unpalatable, and so will not eat enough of it. It may be of low digestibility, so that the animal cannot eat enough to satisfy its needs. Or while not deficient in any nutrient, it may have excess of one or more, such as fat, that will lead to disease.

The usual measures of dietary adequacy are weight gain in young animals, reproductive and lactational performances, and food conversion ratios. All are easy to measure, but reproductive performance may take an inconveniently long time to assess.

Growth rate, more precisely weight gain, is usually measured in animals from weaning until the approach of maturity; that is, during the most rapid period of growth. Daily or less frequent weight gains are measured. If the intake of food is also measured, a relation between weight gain and food consumed may be calculated. This is the food conversion ratio, and is usually expressed as the live-weight gain divided by the dry food consumed.

Lactation makes special demands on the nutrition, even more so in those rodents that, having a post-partum oestrus, may be pregnant while in lactation. A lactating rat may consume her own weight of dry food in 3 days, a mouse in 2; at 21 days the rat may wean a litter two or three times her own weight, and a mouse a litter as much as four or five times her own weight. Measurement of food conversion ratios at this time shows that food conversion ratios of 1 g live weight per 1–2 g dry food consumed are achieved.

Growth immediately after weaning is rapid, and food conversion ratios are likely to be very good at this time. Food conversion efficiency falls off gradually as growth slows down; and in female rats, and possibly in other species, the onset of oestrus is accompanied by a fall in food conversion rate, which recovers once a regular oestrus cycle is established.

The assessment of diet by reproductive performance is the most stringent test of dietary sufficiency, but it may be necessary to go through several litters and two or more generations to obtain significant results. Porter, Lane-Petter and Horne (1963, a) have reported a series of studies in mice, in which the main criterion for assessment was reproductive performance. Two of their observations are of particular interest. First, the same diet formula compounded by three different manufacturers gave grossly different results when fed, thus indicating that a named diet is not necessarily a defined diet. Second, an experiment carried out on outbred mice was repeated on an inbred strain; the results in both experiments were the same, but the inbred strain gave significant results in a much shorter time.

A diet may be adequate when assessed according to growth rate in good conditions, but may fall down when it comes to reproduction. Some workers have recommended different diets for growth and breeding, on the one hand, and for maintenance, on the other. A diet that is adequate in good conditions may fail in bad ones; for example, when the temperature falls, or the animal is stressed. It follows that diets should be assessed in the least favourable conditions, or in conditions of intensive reproduction or at least moderate stress.

Since, apart from specific deficiencies, the adequacy of a diet is only relative, no absolute assessment can be made. Its ability to support growth or reproduction, or to give a good food conversion ratio, must be compared with that of other diets, and on a large enough scale to give convincing results. "A few quick rat feeding experiments carried out over the summer vacation do not constitute adequate testing, nor do longer-term experiments carried out in conditions of undemanding growth and reproduction. Only large-scale long-term trials under intensive conditions, covering at least three or four generations, can carry any sort of conviction that a diet is a good one" (Lane-Petter, W., 1970, d). In the same paper it is reported that four colonies of the same strain of rats were fed on four different diets, but kept in very similar conditions. The main difference between the colonies was in the diet. In one colony a high-fat high-protein diet was fed; at 24 weeks the males weighed, on average, 535 g and the females 350 g. In the other three colonies with adequate but lower calorie diets, the corresponding weights were 435–55 g for males, and 265–305 for females.

V. Sterilization of Diets

Sterilization means the destruction of all living organisms, including the most resistant bacterial spores. There are no degrees of sterilization; no partial or incomplete sterilization. A material has either been rendered sterile or it has not.

There are treatments, however, that kill a large proportion of micro-organisms, but that fail to kill more resistant forms, especially many bacterial spores. Such treatments will destroy all organisms generally associated with epidemic disease in laboratory animals, and are often included in the term pasteurization.

The microbiological standards of laboratory animal diets have been discussed in *Laboratory Animal Handbook 2* (1969). Here it is recommended that, for laboratory rats and mice, *Salmonellae* and *Escherichia coli* Type 1, should be absent, coliforms should not exceed 10 organisms per gramme of diet, and that total viable microorganisms should not exceed 5000 per gramme. It is suggested that treatments severe enough to ensure such standards will be more than sufficient to eliminate arthropods of all kinds, helminths, protozoa and fungi.

Many methods of pelleting diets will in fact expose the mixed ingredients to conditions that will meet these standards, so that the pellet as it emerges from the diet has a very low microbial count. Unless special precautions are taken, however, recontamination may occur. Drepper and Weik (1970) have recommended that nothing less than complete sterility is acceptable in diets for disease-free animals, because of the difficulty of demonstrating that microbial loads are below stated limits. A useful discussion of methods of sterilization will be found in Coates (1970).

Methods of sterilizing and pasteurizing food fall into three groups: heat, chemical methods, and irradiation.

A. STERILIZATION BY HEAT

Bacteria generally are more resistant to dry heat than to moist heat. On the other hand, food materials will oxidize more readily in dry heat, with ample exposure to air, than in an atmosphere that is virtually free of oxygen, such as steam. Dry heat—placing the food in a hot oven—is therefore not a practical method of sterilizing feed; to achieve total sterilization, or even to approach it, will result in too much destruction of the components of the feed.

Moist heat, on the other hand, is much more effective than dry heat in killing microorganisms. The normal method of sterilizing feed by moist heat is by means of an autoclave. In this the feed is placed, the air replaced by steam, and the temperature and pressure raised to given levels and for a given time. Autoclaving will achieve sterilization, with minimal damage to the feed, but some components will be partially destroyed. Among the more heat labile are thiamine, the destruction of which may be well over 50 per cent, vitamin A, and lysine; but other components may also suffer, depending on the exact conditions of autoclaving.

Autoclaving conditions may be anything from 90°C, at subatmospheric pressure, which will achieve pasteurization in 10–20 minutes, to 120°C for up to 30 minutes, or 134°C for up to 7 minutes, both of which can achieve perfect sterilization. A high-vacuum, pulsing autoclave, which can evacuate virtually all the air and replace it by pure steam, will achieve much better penetration of feed pellets than a downward displacement autoclave; and it will also leave the pellets with no higher a moisture content after autoclaving than they had before. Autoclaving pelleted diets frequently makes the pellets stick together in a hard lump, which can only with difficulty be broken up. Pellets containing milk powder or molasses are particularly liable to agglomerate, especially if they are autoclaved in the bag. The difficulty may be partly overcome by spreading the pellets out in a thin layer on wire trays, or dusting them with a chemically inert powder such as talc; but such a powder cannot be considered biologically inert when used in this way.

A special case of heat treatment of feed is by microwave. This is a high-frequency heat treatment which is very penetrating at atmospheric pressure, and it is capable of achieving perfect sterilization. However, there is some destruction of components of the feed, including charring, according to Drepper (1967), who has given a useful summary of methods of sterilization and their effects on feeds.

B. CHEMICAL STERILIZATION

Only one chemical method of sterilization of feed will be considered, namely the use of ethylene oxide. This is a highly reactive and flammable gas, supplied in various mixtures with an inert diluent, such as carbon dioxide. Special equipment is needed for its use: in some cases, a high-vacuum steam autoclave can be modified for the use of either steam or ethylene oxide, at choice.

In some conditions sterilization by ethylene oxide seems to do no more damage to the feed than autoclaving, but there are reports of toxic residues remaining in the feed after treatment (Reyniers, Sacksteder and Ashburn, 1964). This is particularly liable to happen when the cycle of treatment is not exactly according to instruction. A useful summary report of ethylene oxide treatment of feed has been published by Porter and Bleby (1966). See also Charles, Stevenson and Walker (1965); Charles and Walker (1964).

One serious drawback of ethylene oxide sterilization is the time taken. One cycle or, with some night work, two cycles are all that are possible in a 24-hour period, whereas an autoclave cycle need be no more than 20–50 minutes, depending on the size of the load. An ethylene oxide chamber, therefore, needs to be many times the size of an autoclave to handle a similar throughput.

C. IRRADIATION

The only commonly used and effective method of sterilizing feeds by irradiation is by the use of γ-rays; usually from a cobalt-60 source. γ-rays are extremely penetrating, and therefore feed can be placed in containers of convenient size for storage and for the treatment of large quantities.

A γ-ray source is an expensive installation, requiring experienced engineers to operate it. This means that materials to be exposed must be brought to an irradiation plant and taken away again for use. No animal house is likely to be able to afford its own plant, today or in the foreseeable future. The detailed design of the plant being used will determine the size and shape of the packages containing the material to be sterilized.

Useful information about γ-irradiation of diet has been published by Coates, Ford, Gregory and Thompson (1969), by Ley, Bleby, Coates and Paterson (1969), and by Porter and Festing (1970). Almost complete microbial kills can be achieved with a dose of $2 \cdot 5$ Mrad, but at this dose level there will be, rarely, a few survivors. To achieve certain sterility on every occasion, a dose of 5–6 Mrad may be necessary.

At the lower dose, very little damage will be suffered by most diets, but at the higher dose some of the vitamins and amino-acids will undergo change or destruction, to about the same degree as they would if sterilized in a high-vacuum autoclave, or rather less. The physical state of the pelleted diet will not be appreciably altered, in contrast with autoclaving, which may alter it quite markedly.

γ-Irradiation, therefore, has many advantages over autoclaving, even under the best conditions; but it has the great disadvantage of cost. To sterilize a diet by autoclaving may cost perhaps 3–5 per cent of the cost of the diet. To irradiate it may cost 50–200 per cent. If the cost could be substantially reduced, there is little doubt that γ-irradiation would always be the method of choice.

VI. Storage of Feed

Most of the ingredients that go to make up a diet are liable to undergo deterioration on storage. When they are compounded together and pelleted, the rate of deterioration is increased. All diets, therefore, should be used within a short time of compounding and pelleting: the maximum period of storage before consumption should not exceed one month.

Deterioration may be caused by infestation with a variety of arthropod pests, including flour moth, flour beetle, weevils and mites. Fungi and moulds will multiply in the presence of dampness, especially in warm temperatures. Lastly, there is the danger of contamination by wild rats and mice, who will be attracted to food stores and can bring in an endless assortment of infections, many of which will be dangerous.

It is therefore necessary to store feeds in cool, dry vermin-proof rooms. Most feed manufacturers pack their pellets in multilayer paper bags containing 56 lb or 25 kg each. Sometimes there is an inner bag of polyethylene which has been heat sealed and is more or less airtight. All bags to be stored should be unbroken. If the manufacturer can be persuaded to stamp the date of manufacture on each bag, so much the better. In any event, the store of feed should be arranged in such a way that the bags are used in the order of their arrival in the store. It is very disconcerting to discover in the far corner of a food store a number of bags delivered many months ago and forgotten, in which all manner of pest life, not excluding some wild mice, have come to make their home and multiply. And since food is far from cheap, inadequate storage conditions can be costly.

For a general account of insect and acarine pests of stored products, see Munro (1966).

VII. Water

Water is an essential constituent of the diet of all laboratory animals. Species vary in the amount of water they need, in relation to their food intake. A mouse, given food and water *ad lib.*, will drink about the same weight of water as it will eat food. A rat will drink, perhaps, twice as much water as food, weight for weight. A hamster has a low water intake but guinea-pigs and rabbits may be heavy drinkers. It is necessary in all species to see that water is always available.

A. PRESENTATION OF WATER

There are two main ways of offering water to laboratory animals; in an individual container of some sort, or by an automatic drinking device. Individual containers include troughs, fountains and bottles. Open troughs are only suitable for the larger laboratory animals, and for farm animals, and even for these they have several disadvantages. For small rodents they must be regarded as quite obsolete. Fountains are useful for chickens, and they have also been used for guinea-pigs; but like troughs they easily become fouled by the animal, and the water is rendered unfit to drink. For nearly all purposes where an individual water container is chosen, the bottle is used.

1. Water Bottles

A drinking-water bottle delivers water by gravity to the perforated cone or drinking tube attached to its neck. As the animal licks or sucks water from the cone or tube, small bubbles of air enter the bottle to replace the water that is taken out. When there is no licking or sucking, no air enters and no water escapes, unless something goes wrong, in which case the bottle will leak.

Causes of leakage include incorrect dimensions of the hole in the cone, the diameter of the drinking tube or the terminal hole; a badly fitting cone, cap or bung; shaking the bottle—water bottles cannot be used in transit, even for very short journeys; fluctuations of temperature which will alternately contract and expand the air in the bottle, and thus push the water out; and, sometimes, contact between the hole delivering the water with bedding or hair, which may cause the bottle to empty by a siphoning action. Occasionally a water bottle may fail to deliver water because the delivery tube or hole is blocked, by fragments of food or bedding, by calcareous deposits from the water or corrosion of the tube, by an air lock in a tube that is too narrow or at the wrong angle, or by the growth of fungus or alga in the water.

Correct design and use of water bottles, and attention to the quality of the water, should prevent all these difficulties.

Water bottles may vary in size from 100 to 1000 ml. Small bottles need too frequent filling. Large ones are very liable to empty themselves by siphoning when they are half full. For most purposes a convenient range of size is from 200 to 600 ml.

2. Automatic Watering Systems

The alternative to the individual bottle is a water point delivered to each cage, from which the animals can drink at will. This requires a system of piping to serve all the cages, with some type of valve in each cage from which the animals can obtain water, but which will shut itself off completely when not being used by the animal.

The ideal automatic drinking valve can be operated without difficulty by all the animals, at any age at which they may wish to drink, without fail; and it never leaks, or floods the cage. If wire-bottomed cages are used, and the tray underneath drains away, occasional leaks are of little or no importance; but with solid-bottomed cages a leaking valve may mean drowned animals and a flood in the room.

No drinking valve available today can be rated as ideal all the time. Some are unreasonably liable to leak or to blockage, while others have a performance that is nearly perfect. Much depends on the way the system is installed, the design of the valves, the quality of the water and perhaps also the behaviour of the animals. Guinea-pigs in particular will often play with water valves, not to drink, but, it would seem, for fun; and this may result in a considerable flood.

In installing an automatic system, the piping should be as out of the way as possible. Exposed piping collects dust and dirt, which have to be cleaned off: this means extra work. Exposed piping made of plastic is an invitation to an escaped animal to bite through it, with the resultant flood.

There are many designs of valve, some much better than others, Mice and rats lick their water source, so that a water valve for them must require very little force to deliver water. Guinea-pigs and rabbits, on the other hand, suck and their action can be quite violent: they will chew through brass valves, so that stainless steel must be used. Valves for large animals, such as dogs or monkeys, offer few problems. It is important also to ensure that the animal cannot blow water back into the valve and the piping behind it, thus contaminating the water supply.

The quality of the water is often of crucial importance in an automatic system. A water supply with a high or moderate level of dissolved solids— hard water, in short—will certainly deposit some of its solids in the valve, and thus cause it either to fail to deliver water, or to leak. It is often necessary to use demineralized water in an automatic system.

3. Comparison of Water Bottles with an Automatic System

The use of water bottles in an animal house may occupy 10–30 per cent of the labour of animal care. It can therefore be a considerable expense. In many instances, the only reason for bringing animal technicians into the animal house over the weekend is to replenish water bottles. The use of an automatic system would therefore seem to have the great advantages of convenience and of economy. But there are one or two arguments against automatic systems, which ought not to be overlooked.

If a bottle leaks, a cage of animals may be lost, but that is all. If a valve leaks, a cage of animals may be lost, and other cages underneath may be flooded, as well as the floor of the animal room. This entails a lot of work in cleaning up the mess.

Bottles need changing, cleaning and filling. Valves cannot easily be changed or cleaned, which may sometimes be a disadvantage. The piping of valves may entail some work in cleaning.

In many colonies the state of the water bottle gives the animal technician some information about the animals in the cage. A bottle which is not being drunk as rapidly as it should may indicate a sick animal or a blocked tube. Valves do not give this sort of information. It is therefore necessary to look more carefully at the animals when valves are used, and also to check frequently that the valves are working.

For certain purposes, in particular in experimental work, an automatic system is not compatible with the work, and a bottle is necessary. The correct design of bottle can save much work. Perforated cone caps with bayonet fixings are very quick in operation: rubber or plastic stoppers, or screw caps, with glass or metal tubes, are particularly laborious.

There is little doubt that, in large animal installations, where bottles are not essential, the balance of advantage is likely to be in favour of automatic drinking systems; and this is especially true for guinea-pigs and larger animals, and less true for mice, which drink comparatively little. The cost of an automatic system is higher than a supply of bottles, perhaps by a factor of 2. The choice of automatic systems is important, and attention must be given to the quality of the water. If, on the other hand, bottles must be used, some thought should be given to the pattern of bottle and cone or tube, so that a labour-saving choice may be made.

B. WATER AS A VECTOR OF INFECTION

Many bacteria will live, and some will multiply, in tap water, which may also contain fungal spores. When animals drink from an open trough or fountain, they will certainly foul the water, not only with their mouths but also with their feet; and they are likely to scatter bedding and soil into the trough. That is one reason why such methods of giving water are unsatisfactory.

When an animal drinks from a bottle, it will also foul the water, with saliva, with food particles and with any microorganisms that may be in its mouth. However pure the water that is put into the bottle, once the bottle is in the cage it will not be long before it has aquired a generous microflora (germfree conditions excepted, of course) together with enough nutrient material to support rapid multiplication of the organisms. This will rapidly take place at animal-room temperatures, and in a day or two the bottle may be a rich *bouillon*.

If such a bottle is removed and refilled, its microflora will be carried by the aerosol created by the act of uncapping the bottle on to the water tap, and this will infect the next few bottles.

It is possible for an automatic drinking valve to be similarly infected, leading to retrograde contamination of the piping. Most valves, however, seek to overcome this danger, with varying success.

Water must, therefore, be regarded as an important vector of infection in the animal house. Since it is impossible to prevent altogether the contamination of bottles, and of valves, consideration should be given to preventing organisms from growing in the water. This will be considered in the next section.

C. TREATMENT OF WATER

Since water is an important vector of infection, steps have to be taken to ensure that the water supply carries no dangerous organisms, and that infection in the animal house is not readily disseminated by the water. Ideally, sterile water should be supplied, and it should remain sterile until the animals drink it.

Water is used in an animal room, not only for drinking, but also for various washing purposes, including hands and floors. If water is a potential vector of infection, that which the animals drink is likely to represent the greatest danger, but other uses, such as washing, must not be ignored. All water coming into an animal room should be safe.

Water from a municipal main supply, or from a deep artesian well, may have a very low bacterial count on delivery, but if it is kept in an open storage or header tank, the count will increase rapidly. It is wise to assume that all water sources are suspect, and treat that coming into the animal house accordingly. There are several methods, including filtration, heat, ultraviolet irradiation, chlorination and acidification. However it is treated, thought should be given as to whether this may have an effect on the experimental conditions. If the water is not treated in such a way as to make it bacteriostatic, the rapid proliferation of commensal organisms in the water bottle or watering system, except, of course, in the case of germfree systems, should not be overlooked. This is, indeed, a dilemma, for unless water bottles are changed at least twice a day, or automatic watering systems are totally immune to retrograde infection, pure water at animal house temperatures ceases to be pure after a few hours.

1. Filtration

Filtration through a Seitz, millipore or candle filter can remove all particulate matter down to a small fraction of a micron. The finer the filter, the higher the pressure needed to force the water through it. Depending on the amount of particulate matter in the water, the filter will become choked in time and will have to be changed. The advice of a water engineer should be sought, if water filtration is being considered. The use of a Seitz filter will probably be found the most practical and economical method.

2. Heat

Boiling water will kill all microorganisms, with the occasional exception of some resistant bacterial spores. Superheating water will achieve complete sterilization.

There are many systems of heat-treating a water supply, but most depend on filling a closed tank, heating the water by steam coil, and holding it at the desired temperature for a given time, before delivering it. In an animal house there would have to be a battery of such so that cooling could take place before delivery.

This method is somewhat extravagant in use.

3. Ultraviolet Irradiation

Ultraviolet light rapidly kills all microorganisms it strikes, but its penetrative power is very low. A strong source may be unable to penetrate a dust particle, in the shadow of which a microorganism may lurk, and it can only penetrate, in effective amounts, a few centimetres of water. However, there are ultraviolet irradiation plants for treating water, and when properly used they are effective.

4. Chlorination

The commonest and cheapest method of sterilizing water is by chlorination. Free chlorine, usually as sodium hypochlorite, is added to the water, where it reacts with any organic matter, alive or dead. If at the end of the process there is a surplus or residue of free chlorine, sterility may be assumed.

There are available automatic devices for the continuous chlorination of water.

5. Acidification

Other sterilizants than chlorine may be added to water, among them hydrochloric acid (HCl). Strictly speaking this is not a sterilizant in the concentrations generally used; it kills a number of sensitive bacteria, such as *Pseudomonas aeruginosa*, a very common contaminant of water supplies, and it inhibits the growth of a great many more, such as *Escherichia coli*, without killing them. Whereas chlorine is absorbed by organic matter, such as food particles blown into a drinking bottle by the animals, hydrochloric acid is not so affected; it retains its bacteriostatic activity over a wide range of conditions.

Hydrochloric acid is particularly useful for acidifying drinking water, and there is no evidence that in the concentrations used it causes any harm to the animals. It should be added to the drinking water to give a pH of between 2·0 and 3·0. This can be achieved by adding 2 ml of hydrochloric acid (BP) to 3 l of water, or by installing an automatic doser. The resultant acid readily attacks zinc and steel, but not brass, copper, most grades of aluminium or stainless steel. Plastics appear to be unaffected by it.

VIII. Supplementation

From time to time it may be necessary to supplement the food or the water; for example, to feed a substance under test, to make good a nutritional deficiency or to give some medication.

A. EXPERIMENTAL SUPPLEMENTATION

If a substance under test has to be fed to the animal, it may sometimes be convenient to incorporate it in the food. Certain manufacturers of pelleted diets are able to make up diets supplemented in this way. The alternative is to dose each animal individually, or to make up a diet with the test substance added in the laboratory, by the method described on p. 149 or by some other means.

B. MEDICATION

It is sometimes necessary to give a colony of animals some form of medication, such as a coccidiostat or an antibiotic; or to make good a deficiency of a vitamin or some other essential nutrient. The medication may be given to all animals by individual dosing, but in a large colony this will be hardly practical, especially when dosing has to be repeated.

Such medication is usually given either in the food or in the water. Adding it to the food is more difficult, but it can be done as described in the previous section. Many substances can be added to the water; it is only necessary to make sure that this is pharmaceutically feasible, and that the drug or other substance is stable in dilution.

Generally speaking medication has no regular place in laboratory animal care, but from time to time it may be necessary.

CHAPTER 8

Administration and Husbandry

I. Introduction

The operation of any enterprise, be it commercial or a domestic vegetable garden, requires good administration for a successful outcome. This administration necessitates the processes of planning where factors are known and finite, and experiment where factors are unknown. Having put a plan into operation, results require monitoring, feed-back requires assessment and the original plan has to be modified accordingly. The vegetable garden may be planned according to text-book theory, but in practice certain plants may be better suited to one particular area of the plot, or the prevalence of pigeons may eliminate the possibility of growing certain crops without further capital investment.

The idea of administration, in the technical and research environment, frequently produces a hostile reaction on the part of the scientist. This attitude is usually the result of experiencing the not unfamiliar situation where administrative convenience governs the operational procedures. However, the subordination of efficiency because of a dislike of administrative methods is as sensible, to return to the gardening analogy, as sowing all the vegetable seeds at the same time, but the unsatisfactory outcome is tolerated as an escape from the necessity of good planning. This chapter deals with the routine operation of an animal facility and the staff required to carry it out. No excuses are made for the term "administration", since it will be used in the sense of organization and planning in order to satisfy the requirements of the user in the most efficient manner possible within the limits of available facilities and funds. This latter parameter is dealt with at length in the next chapter, where it is stressed that expenditure on animals should be organized so as to keep waste to a minimum; it should also be regarded in the light of the cost of frustrated research resulting from the provision of unsatisfactory animals.

II. Organization

A. THE PYRAMIDAL STRUCTURE

Pyramidal organization is inevitable and operates satisfactorily where functions are laid down and understood. Even in the small animal house a grading of responsibilities supplies incentive to learn and avoids confusion in the absence of any one member. In the larger organization, where administrative efficiency is essential for the prevention of user frustration and economic waste, a carefully thought out reporting structure is vital.

At the top of the pyramid, whether the organization be large or small, there will be a scientist who is experienced in the operation of laboratory animal units and well versed in the use of management techniques. His position within the research organization must be such that his views are respected by the head of the organization and by individual experimenters. The requirements of the latter often produce situations where compromise is necessary within the animal facility; the head of the function must be able to discuss, persuade and make decisions from a position of experience and strength. This niche in the hierarchy is too often filled in an atmosphere of benevolent dispensation, administrative convenience, or even in a spirit of mock malice, as the job that nobody wants to do. There are even instances where no such position exists and the most senior technician reports in theory directly to a professor or divisional head whom he may never see from one month to the next. This latter situation results in each experimenter dealing individually with the head technician, who feels he has not the authority to argue or say "no" and, in trying to please everyone, overstrains the resources at his disposal.

The administration of a laboratory animal facility is not a static process, nor is it efficiently accomplished without technical appreciation and experience. Problems are ever present, even in units operated with enviable success; no sooner is one problem solved than it becomes apparent that two more are clamouring for attention. Because the science of animal management is still a very long way from perfection and because each unit must find its own solution according to its own particular circumstances, the technical management process is a combination of problem solving and improvement. The head of the animal facility must therefore not only possess the necessary experience but should be deeply interested in and involved with the theory and practice of laboratory animal management. A titular head without these attributes is unable to plan successfully, provide technical guidance for his staff or argue convincingly with his colleagues or superiors. Cohen (1960) expands this thesis with reference to an animal facility within a medical school.

That the head of an animal facility is in an unenviable position is an axiom appreciated by those holding such posts and also by those who have successfully avoided the responsibility. Little thanks are expressed when all goes well and problems do not interfere with the course of research, but as soon as difficulties arise the administrative head becomes the reception point for the pent-up frustrations of experimenters. Nevertheless, the successful operation of an animal unit and the challenge of change provide their own rewards.

Having established the qualities of the operational head, the next step is to examine the nature of the supporting staff. Fig. 8.1 provides a basic skeleton organization which can be expanded sideways to suit the extent of the facilities. In a large complex facility it may be advisable to include two chief technicians, but this is not desirable unless the areas of responsibility can be sharply demarcated. Obviously a unit head in a large complex may have much greater responsibility than in a small animal house where he may be responsible for a large number of casual but distinct operations,

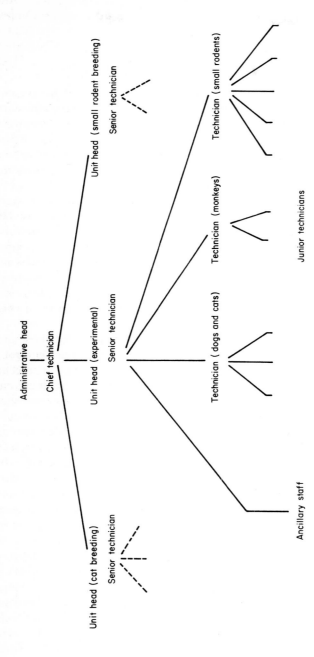

Fig. 8.1. Hypothetical organizational structure for an animal facility.

which may require a lesser degree of administrative control. "Ancillary staff" include those whose functions do not require technical knowledge related to laboratory animals.

B. JOB DESCRIPTIONS AND TITLES

The chief animal technician is frequently the key person in the organization and the success or failure of an operation can depend on the efficiency with which the requirements of that job are met. Primarily the requirement is for a man with wide technical experience in depth, who is also a first-class manager and can understand the problems and requirements of scientists where the supply of animals is concerned. His responsibility is for the day-to-day organization of the animal facilities and for advising the organization head on the practical implications in formulating long- and short-term plans, be they related to the logistics of supply or facility planning. He must be able to identify disease conditions in the animals under his control and be able to propose and carry out suitable corrective procedures. His experience should be such that he can foresee problems resulting from faults in structural design or experimental protocols. In addition, he should possess management qualities which enable him to control the utilization of his staff, institute training programmes and operate his units from the aspect of economic efficiency.

The immediate subordinates of the chief technician are those in charge of specific aspects of the animal operation. They may have control of, for example, guinea-pig breeding, small rodent breeding, maintenance of experimental animals and so forth. In smaller organizations these units can be grouped together for administrative purposes. The unit head should be a technician with at least 5 years' practical experience in the particular field allocated as his responsibility. In some instances, where a new enterprise of a novel kind is undertaken, the unit head will require training and he should be of sufficient calibre that his potential in that field is adequate for the post. As the use of experimental animals increases in quantity and standard, it is from such experienced individuals that the requirement for new chief technicians will be drawn. This aspect will be discussed later in this chapter.

Unit heads carry out their functions through the employment of technicians, junior technicians and ancillary staff. The latter carry out the functions which do not require the expertise of animal technology, such as the operation of sterilizers, incinerators, movement of bulk material, general cleaning and transport of animals to and from laboratories. Where the husbandry of some species requires extensive manual labour, such as cleaning dog runs, such staff can be utilized for these tasks. The animal technician is basically different in that he is required to learn more complex techniques and is expected to identify the unusual on his own initiative. The differentiation between technicians and junior technicians can be a matter of age and experience, or the passing of examinations of the Institute of Animal Technicians.

An efficiently run organization possesses staff who are aware of the nature and limits of their responsibilities, the objectives they are expected to achieve and the purpose of their function within the overall scheme. To have such information down on paper is helpful not only to the individual concerned, but also to his supervisor and colleagues. The documents containing these data are usually referred to as job descriptions. They should be compiled, for preference, for the whole of an operation in one exercise. In that way a correct balance can be maintained and there is less danger of omission or overlap. Individual job descriptions should be compiled for unit heads and above, while generalized job descriptions may be used for technicians, junior technicians and ancillary staff.

The contents and use of job descriptions are described in most manuals dealing with techniques of personnel management and it suffices only to state here the broad outlines. These are:

purpose of the job;
main areas of responsibility;
outline of specific operations to carry out those responsibilities;
identification of supervisor;
enumeration of subordinate staff.

The job description should be finalized after consultation between the person concerned and his supervisor.

The setting of numerically based targets, against which individual achievement is assessed, has little real value and, in fact, is out of context in the laboratory animal operation. The efficiency of an operation is frequently marred by changes in the requirements of research programmes, and the fact that laboratory animal science is subject to a continuous bombardment of unforeseeable factors renders the delineation of success or failure impossible. However, there are aspects which can be subjected to a process of target setting. Even if unattainable, such targets can add interest and satisfaction to an operation and enable a supervisor to identify enthusiasm more easily among his subordinates. Target setting has even been known to convert apathy into enthusiasm. The fields which can be used for this purpose are economic factors, and setting up a specialized facility. The cost of breeding animals is always one which should be kept under close examination and cost targets can be set at the time a budget is compiled or at the beginning of the financial year. Although imponderables, such as manufactured food costs, have a relatively high proportional effect on costs, there are still many aspects which can come under scrutiny and where rational economies can be made. The system for cost control outlined in Chapter 9 provides a good basis for identifying expenditure at a level where savings can be made. It will be found that all levels of staff, if integrated and kept informed within the context of a set target, will respond in a positive manner and will welcome the added interest. The incentive is the satisfaction of performing a task well.

The purpose of job titles is to identify individuals in terms of function and status within an organization. Hence, 'Unit Head, Guinea-pig Breeding' identifies an individual in both these terms. For the outsider, his responsibilities are defined and his position in the hierarchy clarified—thus preventing treading on the toes of his supervisor. For the individual it implies the reaching of a certain level in his career and provides a stimulus to the desire for progression. A decade or more ago, laboratory animal staff tended to be lumped together as a heterogeneous group and regarded as inferior members of the research team—if regarded as members at all. Differentiation was by salary and authority delineated by word of mouth—usually in an incomplete form. This situation is still encountered in many smaller laboratories. The clarification of the staff hierarchy by job descriptions and titles has not only a practical value but also supplies the ingredient of esteem— a vital part of the working situation.

III. Routine Care

A. THE WORK OF ANIMAL TECHNICIANS

The practical details of routine husbandry have been dealt with in Chapter 5. In a discussion on the management of an animal facility it is necessary to consider the broader implications of the work of animal technicians.

The basic work of non-supervisory technicians can be divided into the following categories:

1. Routine feeding, watering and cleaning; management of breeding programmes; maintenance of records; housekeeping.
2. Ancillary tasks, not requiring contact with animals, such as cage cleaning, equipment maintenance, sterilization of bedding and diet, disposal of refuse and carcasses.
3. Experimental procedures and surgical operations. Included in this category are the handling of radioactive isotopes administered to animals, administration of special diets, inoculation of transplantable neoplastic tissue, management of special environments and the production of surgically prepared animals.

A fourth category could also be added which would relate to experimentation in the field of laboratory animal science. This feature will be discussed later.

The second category can and should be performed by unskilled staff, for two reasons. Firstly, the skilled technician commands a higher salary and therefore it makes economic sense for these tasks to be undertaken by non-technical personnel. This arrangement also leads to more time being made available for the acquisition of wider and more detailed technical experience by those capable of accepting a progressive career post. Secondly, these unskilled tasks can be accomplished outside the animal rooms, thus reducing the possibility of the introduction of infective organisms.

The first category is the normally accepted range of duties for the animal technician. Although, superficially, these tasks appear mundane, the success or failure of a unit depends on the degree of expertise used to carry them out. These tasks, although of a routine nature, provide the technician with an opportunity to examine the animals under his care and to identify an abnormal situation. This ability differentiates the experienced, responsibility-minded technician from the novice and is probably the most valuable asset within the animal organization. The capacity for awareness, although produced by training and experience, is only of value if the attitude of the technician is in keeping with its full utilization.

The third category of tasks provides the peripheral but essential component of stimulation of interest. Some of the tasks in this group have been mentioned and from these examples it will be appreciated that a high standard of intelligence and aptitude is required to carry them out. A successful animal operation is dependent on the possession of these qualities by the animal technicians. Too often such tasks are deemed to be too complex for performance by the animal-house staff and laboratory staff are consequently employed to carry them out. In many cases laboratory staff are unable to appreciate changes in health or behaviour of the animals which are readily apparent to those working constantly with them. Most laboratories now employ animal technicians who are quite capable of carrying out these procedures, and it makes economic sense to utilize staff according to the expert knowledge they have. The incorporation of this category into the work of the animal technician also provides a means whereby integration into the research team is facilitated. This aspect is discussed later in more detail.

The work of higher grades of animal staff combines technical expertise with managerial qualities. Neither of these skills is subservient to the other and a successful supervisor requires both. The managerial aspects have already been discussed, but it is relevant at this point to repeat that the efficient operation of a unit is dependent on the best use being made of the potentials of all the personnel involved.

B. RECORD KEEPING

The necessity for maintaining up-to-date, comprehensive records has been stressed in Chapter 5. This aspect of animal husbandry will be considered in some detail, since changes in operational technique should be the result of an intelligent interpretation of well-maintained records and not a subjective synthesis of remembered events.

Records can be divided into three categories:

1. Environmental changes.
2. Breeding data.
3. Health.

1. *Environmental Changes*

It is axiomatic that breeding performance or experimental response can be modified by changes in the environment. Such changes must therefore be monitored and recorded. In this way correlations may be established when the unexpected occurs, and corrective measures taken before catastrophe ensues. The parameters usually measured in animal houses are temperature and humidity. From the aspect of the controlled environment for animal health they are the most important, and while other parameters, such as air changes and movement, air-particle analysis, light intensity and noise levels, may be included from time to time, temperature and humidity should be kept under constant surveillance. The installation of sophisticated air-conditioning plant with humidity control does not solve all problems. Usually some form of compromise must be made, so that under a certain set of circumstances the limits specified may not be achieved. The continuous monitoring of these two variables, by the use of recording thermo-hydrographs, is not usually necessary in all animal rooms, unless wide fluctuations are normally experienced due to the nature of the ventilation system or position of the rooms. However, daily readings from a maximum and minimum thermometer and wet and dry bulb thermometer or hydro-graph are essential in all rooms, and these readings must be recorded—if only to provide an element of confidence for the experimenter. It should be remembered that rooms in the animal house are only inhabited by staff for less than half of the working day and hardly at all over weekends; it is therefore sensible to obtain continuous records in a representative proportion of these rooms. The "Monday morning death" syndrome is frequently the result of modified boiler activity over weekends. While a group of small rodents housed in plastic cages can modify their microclimate to some extent, the long-term experimental animal, housed in cages with grid floors, is much more at the mercy of environmental changes. Rooms which house such long-term animals should always be equipped with continuous-recording systems.

An essential feature of the modern animal facility is the positive air pressure maintained in animal rooms. The maintenance of such a pressure can be very simply monitored by the introduction of a small copper tube in the dividing wall between the room and corridor. The differential pressure can be quickly checked and recorded by applying a suitable water mano-meter to the corridor end of the tube.

2. Breeding Data

The necessity for maintaining adequate breeding records has been stressed in Chapter 5. Such records are essential for the purposes of breeder selection and early elimination of below-standard breeders, of background data for certain experimental purposes, and as an early warning of the presence of disease. In the breeding of costly species such as cats and dogs, accurate records are essential for making meaningful forecasts. Such records also provide data on seasonal fluctuations, which may be vital where holding facilities are small and demand is variable.

These data also provide a means whereby the efficiency of a unit can be measured against that of others. Since operational excellence depends on a spirit of continuous improvement, such comparative data can provide the required impetus.

For the small rodent breeding unit, where the life of the breeder is measured in months rather than years, record keeping is a comparatively simple operation. Where such breeding is contained in individual rooms (as described in Chapter 5), which are operated on the basis of complete stocking-up and running-down cycles, the data may be entered into suitably ruled exercise books. These have the advantages of being robust and provide easy storage. Individual female breeders may be identified by ear markings and entered in numerical order for easy reference. A master reference book should be maintained in which the numerical system is broken down into genealogical relationships. Basic data to be recorded should relate to breeder details of date of birth and date first mated followed by sequential litter data. The latter should include:

1. Male code number.
2. Date mated or returned to mating cage.
3. Date when litter born.
4. Number born.
5. Number weaned.

When culling of a surplus sex takes place before weaning, or cross-fostering is employed, this information should be recorded in order to identify those females with a below-standard record of postnatal care.

The recording of breeding data from the larger species—cats and dogs—is more complex since the results may show considerable variations for each individual breeder. The breeding of these species is much less of a science than an art, since the factors producing such variability are largely undetermined. It is therefore essential that detailed records be kept for the greater understanding of the potential of a colony and to enable correlations to be established and suitable action taken. Since very large numbers of breeders are not usually involved in these species, a card index system can usually be employed. A master card for each female breeder can be backed up with a card for each litter produced. The loss of a single litter or puppy constitutes a substantial financial loss and therefore the details surrounding mortality in these species must be as comprehensive as possible. Progress in breeding efficiency can only be achieved by frequent reviews of data in order to establish the prime causes of mortality, failure to become pregnant, small litter sizes, poor survival rates and so forth. Additional data include vaccination and inoculation dates, litter growth rates, dates when dietary or environmental changes occur. The importance of identifying the main factors influencing breeding performance, in species having infrequent litters with small average litter sizes, is obvious if an economical operation is to be maintained. The identification of such factors can only be made where detailed, relevant records are kept.

3. Health

The necessity for maintaining records relating to animal health has been stressed in Chapter 5. The reasons are associated with the prevention of disease outbreaks and a thorough familiarity with the experimental animal, whether user-bred or purchased. Such records fall into three categories—post-mortem results, bacteriological results and observations of abnormal conditions not necessarily affecting the health of the animal. These records can conveniently be maintained in separate books, but frequent analysis and summaries should be made. Where this is neglected, the entering of an event in the appropriate book becomes an end in itself and no useful purpose is served. The aim of records is to establish correlations with other events, to anticipate an extension of a disease condition and to record the incidence of such conditions as chromodachryorrhea and alopecia; the periodic summarizing of such data provides a convenient way of examining the progress of a breeding colony and providing background data for long-term experimentation, as in toxicological work.

In breeding colonies of species such as cats and dogs, the correlation of health records with those relating to environment and breeding results becomes an essential feature of an efficient operation.

No attempt will be made here to describe recording methods in detail, since this depends so much on the nature of the colony, the purpose of the operation and the personal preferences of those in administrative control.

IV. Selection of Staff

It is apparent from the previous chapters that the management of an animal facility, although in some measure a defined technical operation, still leaves scope for the exercise of experimentation and craftsmanship. The strict observance of rules and scientific method will not necessarily result in a successfully run animal house. The human factor is the key. The qualities required of supervisory staff in the animal house have already been discussed; consideration is now given to the qualities required of the lower grades. These fall into three categories: aptitude, ability to acquire knowledge, and experience.

A. APTITUDE

This quality is a fundamental requirement. Where present it can be developed and utilized; where absent it cannot be created. It can be expressed as a love for and working with animals, a devotion for the well-being of animals and an inherent understanding of their requirements.

Many laboratories in the past—and, sadly, many to this day—equate the animal technicians with porters or labourers, assuming that the day-to-day operations are non-technical, mundane and requiring little skill. This attitude is a reflection of the importance placed on the quality of the animal utilized. The work of animal house staff involves weekend work, which must be carried out with the same degree of conscientious thoroughness as during normal weekday working hours. When an emergency arises they are required to return to the animals under their care at any time of the day or night. Breeding bitches have a habit of whelping in the small hours of the morning; ventilation and heating equipment choose weekends as the time to go wrong; experimental requirements often necessitate the employment of animal house staff outside normal working hours. To fulfil his function to his own and his employer's satisfaction the animal technician must regard his job in the light of a devotion to his task. Such a standard is high, but is still a basic requirement. Without such an attitude the technician will lose interest in his job and be unable to fulfil his task to an adequate standard.

The possession of the right temperament for working with animals is difficult to identify during an interview but becomes apparent very soon after employment. Some animals, like some employers or employees, are easy to work with, others require a great deal of patience. This does not imply that a placid temperament is necessarily required, but rather an understanding that animal behaviour can be erratic and is not personally directed against the technician. A sense of responsibility must be developed and a realization that the welfare of the animals under his care is dependent on his own attitude as a "superior being" capable of inflicting suffering or ensuring well-being.

The employment of staff without this inherent attitude results not only in a lowering of the standard in the animal house but also provokes antagonism on the part of other members of the staff. The staff of an animal house is usually a closely knit group of people with a similar attitude to their work, and the introduction of a member lacking the same attitude can rapidly provoke staff discord. It is important, because of the necessity for fostering a team spirit, that the attitude be right and the new employee be temperamentally compatible with the existing staff.

B. THE ABILITY TO ACQUIRE KNOWLEDGE

The degree of understanding required of a competent technician is high. The work of the animal technician has been shown to equate with the technical level of the laboratory technician. The term "animal technician" has been used to identify persons who have to carry out duties requiring technical skills and the ability to interpret novel events in the light of those skills. The ability to undertake experimentation and maintain records must also be present. The term technician is used in Great Britain to denote a person who is capable of learning and carrying out technical operations in support of a graduate scientific staff. In the field of animal technology the animal technician rightly deserves his title. The separation of technical staff from those undertaking manual duties has already been touched on.

What background qualifications are necessary, then, for the untrained animal technician? While basic educational standards are useful as a guide to intelligence and the potential of an individual for learning, an inherent interest in animals and a desire to learn are also important. The depth of understanding increases with the advancement of laboratory animal science, and the selection of staff must relate to the degree of understanding required. Nevertheless, the operations of the technical staff are such that the aptitude for understanding the meaning of technical terms and basic biological principles is essential. The subject of staff training is dealt with in detail later in this chapter, where it is shown that formal training can reach a very high standard. The ultimate standard is not necessarily required of all staff and recruitment should be approached with the correct balance in mind. While a basic standard of intelligence is desirable, too rigid an attitude towards formal qualifications may exclude potentially excellent technicians who excel in aptitude and attitude. The decision where to draw the line depends on particular circumstances; the section on training, later in this chapter, may assist supervisors in making this decision.

C. EXPERIENCE

While relevant experience is an essential requirement for higher grades of staff, where a new facility is to be developed it may be a doubtful asset in junior employees. So much depends on where the experience was gained. The acquisition of someone else's bad habits is more difficult to erase than *de novo* training in correct concepts and techniques. This is not to say that a background of experience in a technician able to utilize such knowledge in a different environment is not of value. Adaptability is a considerable asset, since no two animal houses are or can be operated in the same way, due to variations in the species and qualities required and the facilities available.

Experience need not be related to technical expertise. The technician with experience and a record of success in breeding and maintaining small rodents may have excellent qualities as a staff supervisor and unit manager. These assets could be well utilized if such a person were to undertake the setting up of, for instance, a guinea-pig and rabbit breeding facility.

The introduction of experienced staff into an existing unit can have benefits provided the supervisor is prepared to be objective about such experience where it differs from existing practice. Laboratory animal science is a discipline undergoing continual change and improvement; a rigidity in approach on the part of supervisors benefits neither the discipline nor the unit where it is practised.

Where the nature of the vacant post requires a high standard of technical expertise and promotion from within is not possible, the possession of formal qualifications of the Institute of Animal Technicians at a suitable level provides an assurance that the necessary knowledge has been acquired.

Finally, it must be stressed that practical expertise in handling a variety of laboratory species takes many years to acquire. Much of this expertise cannot be gleaned through books and the technician possessing such experience represents a very valuable asset to the laboratory in which he is employed—a fact too often appreciated only when replacement recruitment is undertaken.

D. GENERAL CONSIDERATIONS

Preconceived concepts play a large part in delineating the requirements for any post. The selection of animal technicians is no exception. Age and sex are two factors which play a large part in making decisions before selecting a candidate for employment. Where the requirement is for an inexperienced junior technician as a trainee, the temptation is to consider only school-leavers or persons up to the age of 20. In practice, many women in their thirties or in middle-age make excellent animal technicians from the aspects of technical ability, reliability and conscientiousness. It is often considered that women are better than men at handling small rodents while men are better with the larger species. This is possibly true to a large extent, but the truth lies in the physical requirements of the job rather than in some indefinable affinity. Cage sizes required in the husbandry of guinea-pigs and rabbits often dictate that male technicians should be employed; the husbandry of dogs requires a physical effort that can be more easily sustained by male staff. Where such physical requirements are not present, male or female staff may equally well be employed.

Employment in an animal house used to be regarded as a dead-end job equivalent to a labourer or caretaker. The technical demands of these positions in the present day have dictated a worthwhile career structure for the intelligent non-graduate. The career-minded young animal technician may be assured of a comparable salary to his colleagues working in the laboratory and far greater opportunity for supervisory responsibility at an early age.

V. Training of Animal Technicians

Training is of two kinds: practical or on-the-job and formal or tutorial. Practical training is undertaken through day to day supervision and experience gained in carrying out duties. Formal training consists of lectures, demonstrations and tutorials with practical tasks associated with such teaching. At the present time such formal training is only available in Great Britain on a part-time basis, the student normally being employed at the same time in the capacity of an animal technician. The two aspects of training are therefore given concurrently and, because of the essentially practical nature of the expertise, this is as it should be.

It cannot be too strongly emphasized that adequate planned training is indispensable in this field. However sophisticated the equipment and facilities, untrained or unsuitable personnel cannot be expected to produce and care for satisfactory laboratory animals. On the other hand, well-trained animal technicians can often provide excellent animals under conditions which fall short of ideal.

A. ON-THE-JOB TRAINING

Practical training, during the course of employment, is all too often undertaken in a haphazard manner. Because the nature of the day to day operations appears superficially mundane, the training of new staff may be less formal than is desirable. The technician in charge of a unit is the person responsible for this aspect of training and a formal approach is helpful to himself and trainees. It is convenient to separate two grades of basic training—elementary and advanced—and for the unit head to compile a syllabus for each stage which is relevant to his own particular unit.

The elementary stage should cover all aspects of basic training to enable a junior technician to accomplish his task as laid down in his job description. The following list can be taken as a guide to such a syllabus:

1. Cage cleaning.
   Method, frequency, replacement.
2. Feeding.
   Type of diet, frequency of administration, method of preparation, disposal of unused food, container cleaning and replacement.
3. Bedding.
   Type, frequency of replacement.
4. Environment cleaning and maintenance (racks and room).
   Methods, frequency.
5. Animal handling.
   Techniques under different conditions during normal breeding cycles, method of weighing, painless killing techniques, holding animals for inoculation.

   6. Breeding technique.
   7. Record keeping.
   8. Sterilization.
      Methods, frequency.
   9. Routines outside normal working hours.
  10. Elementary clinical observations.
  11. Fire precautions and escape method.
  12. Personal hygiene.
      Discipline on entering and leaving animal area, protective cloth-
      ing, reporting personal illnesses or those of domestic pets.

The unit head should satisfy himself that such training is properly carried
out by himself or a senior technician and that the trainee has absorbed the
information implanted. Periodic informal oral examination will satisfy the
supervisor of progress being made and provide the trainee with confidence.
Setting out the outlines of such an elementary syllabus on paper will enable
the trainee to seek information on topics in which he feels uncertain.

In practice, it takes from 1 to 2 years before an inexperienced technician
acquires complete confidence in what he is doing and what is expected of
him.

The stage of more advanced training should also be placed on a formal
basis. In this way the unit head can keep a record of who has been trained
in which technique, and can plan accordingly. The trainee becomes aware
of the extent and interest of the job and will feel that his career has not
levelled out purely into a routine.

The syllabus for such advanced training must depend on the nature of
the unit and the facilities available. However, even the most basic of units
should have scope for developing the expertise of the technical staff. As a
guide to such a syllabus the following ideas are presented as a stimulus to
thought.

   1. Clinical diagnosis.
      Measurement of body temperature, pulse and respiration. External
      appearance of eyes, hair and skin. Discharge from eyes and nostrils.
      Examination of faeces. Behavioural observations.

While in the larger species, such as cats and dogs, final diagnosis should be
made by the veterinary surgeon, the recording of symptoms by a trained
technician over a period of time can provide valuable data for such a diagnosis. In the more common laboratory species the experienced and well-
trained technician is sometimes in the best position for making a diagnosis.
In any event the routine recording of variations from the normal condition
is an essential part of the technician's task.

   2. Germfree techniques.
      Theory and construction of germfree isolators. Hysterectomy

technique. Management of animals in an isolator unit. Passage
techniques in and out of the isolator. Total sterilization methods.
Observation of germfree animals in association with known conditions resulting from the germfree state.
3. Operations not requiring a Home Office licence.
   Prophylactic inoculations, administration of antibiotics, administration of gut flora complex to germfree animals, taking vaginal
   smears.
4. Operations requiring a Home Office licence.
   Thyroidectomy, hypophysectomy, removal of body fluids,
   administration of substances by subcutaneous, intraperitoneal or
   intravenous routes.
5. Laboratory sterile methods.
   Taking samples under sterile conditions from live animals or
   cadavers for microbiological testing.
6. Records.
   Techniques for the analysis and summarizing of records and testing
   for correlations.
7. Expenditure and stock control.
   Elementary cost recording and techniques for stock disposal and
   replacement.

It must be repeated that this list is not intended to be comprehensive,
but is only presented as a guide. Additional aspects can be introduced where
special facilities are available or requirements demand. Whatever the final
syllabus, the unit head should arrange a training programme which suits
the time available and the abilities of his staff. The effect of such training
should be assessed by practical and oral examination on an informal basis.
In many cases, where formal external training, as described in the next
section, is not practicable, this advanced training undertaken in the course
of a technician's employment should be put on a more formal footing and
supplemented by recommended reading of the theoretical aspects.

A technician who is fully and competently conversant with the advanced
stage of on-the-job training is an asset not only to the animal division but
also to laboratory staff. The latter are rarely trained in the basic techniques
of animal husbandry and handling, and the experienced animal technician
can himself perform a vital training task.

The training of staff below the level of unit head, or the senior technician
in charge of the animal house, does not represent a climax. Much of the
contents of this book have stressed the need for innovation and experimentation in laboratory animal science and continuous progress being
made. As in other technical disciplines the need for training, experimentation and keeping abreast of developments by other units is a responsibility
to be undertaken by senior staff, including the graduate in administrative

control. The encouragement of an experimental approach, undertaken along sound scientific lines, will reap practical as well as psychological rewards.

B. FORMAL TRAINING

Formal tutorial training is best arranged in institutes under the auspices of local education authorities. Such courses are arranged in collaboration with the examining body, the Institute of Animal Technicians, and take the form of evening classes or combined day-release and evening classes. It is essential that such training be undertaken concurrently with practical on-the-job training by those already employed as animal technicians. Lectures on the theoretical and laboratory aspects should be given by graduates in a biological discipline and practical demonstrations by very experienced senior animal technicians. Such courses should be devised to conform to the syllabuses of the Institute of Animal Technicians.

In 1950 the Animal Technicians Association was founded, which changed its name to the Institute of Animal Technicians in 1965. The main purpose of this Association was and is to concentrate on the education and training of animal technicians in order to meet the developing and exacting requirements of laboratory animal science. In this task it has been extremely successful and has achieved much in raising the standard and quality of laboratory animals on which successful biological research depends.

The Institute examinations are based on academic knowledge and practical competence. The Intermediate examination may be taken by candidates who have experience in the care of laboratory animals over a period of not less than 2 years and are employed in an approved laboratory; or who have not less than 1 year's such experience and who hold the GCE at "O" level in English Language and either Biology, Human Physiology or Zoology. Registered Animal Nursing Auxiliaries may take this examination without the preceding requirements. The Associateship of the Institute may be obtained by examination after a minimum of 4 years' experience in laboratory animal care, or 3 years provided certain academic qualifications have been obtained; Registered Animal Nursing Auxiliaries require 2 years' relevant experience and in all cases the Intermediate Certificate must be held.

The Fellowship of the Institute represents a stage of advanced practical competence and detailed appreciation of the technical principles relating to successful laboratory animal management. This stage is reached either by examination or by thesis and candidates must have held the Associateship Diploma for at least 2 years.

The requirements for entrance to the examinations are thus based not only on academic level but on experience gained in practical application. This emphasis on experience is an important theme in this technical field and a high degree of competence in practical work is regarded as an essential for successful examination candidates.

In many instances, due to geographical location or inadequate numbers, it is not possible for technicians to attend formal courses of training at technical institutes; reliance must then be placed on the training support of senior technical and graduate staff. The syllabuses of the three levels of examination supervised by the Institute are outlined below, in order to assist the organization of formal training courses and to provide for prospective employers a guide to the technical level achieved by candidates for employment. Entrance to the Institute examinations is permissible without formal training at a recognized institute.

INTERMEDIATE SYLLABUS

1. *Elementary mammalian anatomy and the basic principles of physiology, with special reference to the rat and the rabbit.*
   Circulatory system, including structure and functions of the blood; respiratory system; digestive system; urogenital system.

2. *Simple arithmetic.*
   Conversions between the metric and other systems in common use.
   An understanding of the Centigrade and Fahrenheit scales and conversions between the two scales.
   Simple thermometry, to include the maximum-minimum and clinical thermometers; bimetallic strips; thermocouples.
   Calculation of percentages.
   Simple proportion.
   Measurement of area and volume.
   Preparation of solutions of known strength.
   Costing of foodstuffs.

3. *The Cruelty to Animals Act, 1876.*
   An understanding of the legal requirements covering the management of experimental animals.
   Licences and certificates.
   The Conditions attached to every licence.
   Administration of the Act, including visits of Home Office Inspectors; recording of experiments; visitors to experimental areas.

4. *Hygiene.*
   Disposal of used bedding and carcases.
   Personal hygiene.
   Methods of sterilization; principles and use of washing machines; free-steam chests; autoclaves; hot air ovens.
   Classification and use of disinfectants.

Use of fumigants.

Checks for efficiency of sterilization processes, e.g. Browne's tubes; autoclave tape; spore strip.

5. *Nutrition.*

Diets suitable for the common species, including foodstuffs which could be used in an emergency.

Correct storage of foodstuffs.

Detection of deterioration and infestation of foodstuffs.

Functions of protein, carbohydrate, fat, mineral salts and accessory food factors, with special reference to the need for a dietary source of vitamin C for the guinea-pig and monkey.

6. *Handling and sexing of common laboratory animals.*

Mouse, rat, guinea-pig, rabbit, Syrian hamster, ferret, cat and dog.

NB When handling an animal candidates should also be able to give an indication of the animal's age and body weight.

7. *Routine care of common species and of the Rhesus monkey.*

Methods of feeding and watering.

Bedding and nesting materials.

Cleaning of cages, equipment and premises.

Environmental temperature and humidity, and the optima recommended for various species.

Inspections for injuries, including sore hocks and overgrown teeth and claws.

Normal body temperatures.

8. *Ill-health in the common species.*

An elementary knowledge of the causes of ill-health.

Recognition of the signs of ill-health, e.g. loss of condition, respiratory infections, infestations.

9. *Breeding of the common species.*

Recognition of good breeding animals.

Elementary knowledge of oestrous cycles.

Gestation periods.

Average litter sizes.

Average birth, weaning and adult body weights.

Selection of good breeding stock.

Duration of economic breeding life.

Breeding systems: matings at post-partum oestrus; monogamous pairs and harems; closed colonies; inbreeding and random breeding.

10. *Methods of identification for common species.*

11. *Euthanasia for common species.*

12. *The use and care of animal and food balances.*
    Elementary principles of spring, beam and torsion balances.

    NB The handling and sexing of laboratory animals is a compulsory section of all examinations.

ASSOCIATESHIP SYLLABUS

1. *Mammalian anatomy and physiology.*
   Introduction to skeletal, cardio-vascular and endocrine systems and organs of special sense.

2. *Elementary microbiology.*
   A sufficient knowledge of microbiology for the management of SPF colonies and isolation units for infective animals; an understanding of the problems of sterilization.
   Elementary microscopy.

3. *The Cruelty to Animals Act, 1876.*
   Objects and scope of the Act.
   Procedure for obtaining licences and certificates and completing forms of application.
   Recording of experiments and preparation of Annual Returns of Experiments.
   The Conclusions and Recommendations of the Departmental Committee on Experiments on Animals. (Littlewood Committee)

4. *Nutrition.*
   Chemical composition of foodstuffs.
   Composition of cubed and pelleted diets in general use.
   Nutritional requirements of laboratory animals.
   Nutritive value of common foodstuffs.

5. *Hygiene.*
   Uses and properties of disinfectants and antiseptics.
   Factors governing the choice of method of sterilization, disinfection and fumigation.
   First aid in the animal house and special precautions to be taken when handling animals.

6. *Disease and infestation.*
   Recognition and control of common diseases of microbiological and parasitological origin of laboratory animals.
   Identification and simple life-cycles of common pests, e.g. bed-

bugs, lice, fleas, ticks, mites, flies, beetles, flour-moth, weevil, food-mites, wild rodents.
Detection and control of infestations.

7. *Management of isolation units for infective animals.*
Importance of strict discipline and routine.
Handling of infective animals.
Precautions against contamination and cross-infection.
Disposal of infected materials and carcases.
Disinfection of isolation units.
Preparation of infective animals for post-mortem examination.
Management of post-mortem rooms and equipment.

8. *Manipulations and techniques.*
Routes for injections and methods used for the withdrawal of body
    fluids.
Preparation of animals and equipment for aseptic procedures.
Administration of anaesthetics.
Post-operative care.
Techniques for taking vaginal smears and recognition of the phases.
Techniques for passing stomach tubes.
Collection of specimens, e.g. urine, faeces, blood, blood smears.
Preparation of animals for post-mortem examination.

9. *Breeding of laboratory animals.*
Additional species to be studied:
    gerbils, Chinese hamster, cotton rat, common laboratory primates,
    duck, chicken, turkey.
Planning and administration of breeding programmes.
Record keeping.
Inbreeding techniques.

10. *Farm animals used in the laboratory.*
Elementary knowledge of the breeding and maintenance of the
    horse, cow, sheep, goat and pig.
Recognition of the signs of ill-health in these species.

11. *Other species.*
Elementary knowledge of the maintenance of frogs, toads and fish.
Recognition of the signs of ill-health in these species.

12. *SPF (Specified Pathogen-Free) techniques.*
Design of SPF units.
Sterilization of building.
Sterilization and introduction of supplies.
Techniques for obtaining and rearing SPF animals.
Transport containers for animals.

Effect of sterilization (or other treatment) on diet.
Special staffing requirements.
Principles of routine microbiological testing.

13. *Animal house equipment.*
Caging and racking for laboratory animals.
Use and care of labour saving devices.
Use and care of special equipment, e.g. animal clippers; marking equipment.

14. *Animal house management.*
Work schedules and deployment of staff.
General maintenance and repairs.
Sources of animals, foodstuffs, equipment.
Despatch, transport and receipt of animals.
Isolation and quarantine of animals.
Functions of the Laboratory Animals Centre.
Sources of information about laboratory animals.

15. Subjects covered by the Intermediate syllabus not specified in this syllabus.

   NB The handling and sexing of laboratory animals is a compulsory section in all examinations.

FELLOWSHIP SYLLABUS

1. *General anatomy and physiology of laboratory animals, with special reference to mammals and birds.*
Elementary embryology.
Structure and functions of the skeletal, vascular, respiratory, digestive, urogenital, endocrine and nervous systems.
Haematology: routine techniques, e.g. blood counts, blood smears.
Histology: structure and function of the cell:
   simple preparation of permanent sections of tissues, e.g. epithelium, connective tissue, muscle, nerve;
   elementary histology of organs;
   normal post-mortem appearances.
Chemical pathology: impaired liver function:
   impaired renal function.
Immunology: allergy and immunity;
   other defence mechanisms against disease.
Microbiology: classification of the main groups of microorganisms:
   simple culture techniques for bacteria;
   examination of organisms.

2. *Laws relating to the use of laboratory and farm animals.*
   Cruelty to Animals Act. 1876.
   Relevant parts of:
   Protection of Animals Acts 1911 and 1934.
   Protection of Animals (Anaesthetics) Acts 1954 and 1964.
   Dogs Act 1906.
   Veterinary Surgeons Act 1948.
   Laws relating to farm animals including licensing and transit.
   Laws relating to importation and exportation of laboratory and
   farm animals.
   Rabies Order of 1938 *and* The Exotic Animals (Importation)
   Order 1969.

3. *Nutrition.*
   Preparation of special diets and associated feeding techniques.
   Diet manufacture.
   Calculation of nutritive value of diets.

4. *Elementary genetics.*
   Mitosis and meiosis.
   Genes and chromosomes:
   haploid and diploid numbers; alleles; dominant and recessive
   genes; homozygosity and heterozygosity.
   Inheritance of characters controlled by a single gene.
   Segregation of genes carried on separate chromosomes.
   Segregation of linked genes.
   Sex chromosomes; sex linkage.
   Back-crossing.
   Mutation.
   Gene interaction.
   Genotype and phenotype.
   Quantitative inheritance.
   Inbreeding and random breeding; inbreeding depression; hybrid
   vigour.

5. *Disease.*
   A knowledge of the transmissible diseases and infestations of common species. causes; signs; methods of control and elimination,
   including prophylaxis.

6. *Production and management of gnotobiotes.*
   Design of isolators.
   Sterilization of isolators.
   Sterilization and introduction of supplies.
   Techniques for obtaining and rearing gnotobiotes.
   Transferring animals between isolators. including the use of trans-
   port containers.

Special dietary requirements.
Characteristics of germfree animals.
Principles of routine microbiological testing.

7. *Radioactivity in animal experiments.*
   Types of hazards of atomic radiations.
   Dose units and maximum permissible radiation levels.
   Relative toxicity and permissible levels of commonly used isotopes.
   Monitoring methods.
   Handling, caging and maintenance of radioactive animals.
   Decontamination of cages, equipment and personnel.
   Disposal of radioactive cadavers and materials.
   The principle requirements of the Radioactive Substances Act, 1960.

8. General knowledge of the budgerigar, canary, pigeon, carp, trout, newt, salamander and grass snake.
   Breeding of frogs and toads.

9. *Animal house design.*
   Structure and choice of materials.
   Layout of rooms.
   Ventilation and control of temperature and humidity.
   Heating, lighting and insulation.
   Supply of other services.
   Animals houses for special purposes, e.g. SPF, radioactive isotope usage.

10. *Administration.*
    Book-keeping and costing for staff, animals, equipment and food.
    Elementary statistics.
    Management of staff.

11. First aid in laboratory hazards.
    Zoonoses.
    Precautions with toxic substances.

12. Subjects covered by the Intermediate and Associateship syllabuses not specified in this syllabus.

    NB The handling and sexing of laboratory animals is a compulsory section in all examinations.

These syllabuses apply to the academic year 1971/72.

The chief advantages of such formal training are the provision of a series of goals providing the young technician with a career sense, the realization of the wider aspects of the job, and providing both employer and employee with confidence in carrying out the full range of duties falling within the animal technician's sphere.

C. JOB ROTATION

The tendency in any technical field to become overspecialized is a difficult one to resist. A case can be made out for narrow specialization where the field is large and the technical requirements of certain specific aspects are such that efficiency can only be obtained by such specialization. In the field of laboratory animal science, specialization is concentrated on the requirements of individual or closely related species. Thus a technician may regard himself as a specialist in small rodents, rabbits and guinea-pigs, dogs and so forth. While much is still to be learned from experimental husbandry and management of laboratory species, and this advancement in knowledge is best undertaken by the specialist, there is a greater need for an expansion of experience on the part of the career technician. The emphasis on the use of particular species within a laboratory may change—sometimes gradually, sometimes from one day to the next. The senior technical staff must have the basic knowledge and experience to advise on the implications of such changes and to take the necessary steps in a competent manner.

In the large research organization such training can be undertaken by rotation between units having different functions. The ideal career stage for such training is when a technician has achieved the status of senior technician or technician by virtue of experience and successful completion of formal and on-the-job training. Rotation is often left too late and a laboratory may find itself in the position of having highly specialized unit heads with little or no experience of operations apart from their own. At this level rotation may result in a drop in unit efficiency. At the lower level of junior technician, rotation is not advisable due to incomplete on-the-job training in a particular function, or lack of interest in other species or job progression. The advantages of job rotation, in the larger laboratory, accrue to both employer and employee and can be accomplished with the minimum of disruption. To the employer it provides flexibility by having a pool of senior technicians with the capacity to undertake different jobs in a competent manner in the event of an emergency, or by internal promotion to fill a higher post as a result of staff wastage. To the career technician it provides a wider field of interest with greater opportunities for promotion.

In the smaller animal division, often devoted only to the breeding and maintenance of two species, the opportunities for acquiring wider experience are more limited. In these cases much can be accomplished by an enlightened management prepared to accept temporary inconvenience. Senior technicians can be exchanged between laboratories able to offer training in a different expertise. Problems over security can arise with commercial undertakings, but the smaller specialized laboratories, requiring the interchange of technicians, are usually to be found in universities and public or charitable research laboratories where such considerations do not apply.

VI. The Working Environment

It can be said that a research director gets not only the animals but the technicians he deserves. For the efficient management of experimental animal breeding and maintenance, the technical staff must not only be carefully selected before employment but must be given a working environment conducive to job satisfaction. This implies not only acceptable physical conditions but also the satisfaction of training and the feeling of operating within a team. The first two considerations have already been discussed in this and previous chapters. We are concerned here with the relationship of the animal technician with colleagues in the animal house and the laboratory.

It has been shown in previous chapters that good production techniques and considerations of disease prevention are tending to reduce the size of animal rooms. Large areas for rodent production and large multi-purpose rooms for the maintenance of experimental animals have been subject to difficulties in environmental control and rapid spread of infections. The ideal of one room/one technician, with the attendant advantages of individual responsibility and personal pride in a specific area, is not confined to small rodent breeding but may be extended to the management of a cat pagoda or an individual experimental room. For the technician who has no ambitions to take on wider responsibilities, this system does much to maintain interest and provide a measure of congenial competition. It also simplifies the task of the supervisory technician who is then able to identify and rectify faults in technique. This approach, however, implies solitary working conditions which do not necessarily appeal to all technicians. It is usually found that older staff and male technicians prefer to work on their own, while younger female staff prefer working in a group or in pairs. The one room/one technician concept can usually be modified to a plan of working involving two technicians/two rooms. The limiting factors are the sizes of rooms, the facilities for outer clothing changes and the nature of the husbandry involved. Where an on-the-job training programme is involved, an experienced technician can work with the trainee. The organization of such working schedules must, in any event, be arranged according to the personalities of the individuals concerned.

The relationship between the animal division staff and the laboratory worker is of prime importance in the conduct of animal experimentation. The success or failure of such a relationship depends mainly on the attitude of the experimenter. Where animal technicians are regarded as low-grade manual workers, it is hardly surprising that little or no relationship exists. The validity of experimental results on living animals depends to a large extent on the quality of those animals and the manner in which they are maintained before and during the experimental period. The animal technician must be appreciated and used in the context of his training, experience and expertise. Such knowledge is of prime importance to the research worker, who usually lacks this detailed experience. The experienced animal technician should be incorporated into the research team on the same basis as any other technician possessing specialist knowledge. Not only is he in a better position to observe clinical or behavioural changes at an earlier stage, but advice and opinion can be expressed during the planning phase of experimentation. The purpose of animal technician training is not only to provide for efficiency in matters of husbandry, but to appreciate the purpose and effects of experimentation. The practical nature of this training also enables him to carry out certain experimental procedures, in which he is often more capable than laboratory staff, by virtue of his familiarity with animal handling.

This concept of the complete animal technician is becoming more accepted and is very successful where put into practice. It involves careful manpower planning and has advantages in making the best use of the technical expertise available in the animal house and the laboratory. The integration of the animal technician into the objective of his prime function provides a sense of purpose which maintains his interest and acts as a catalyst for the development of his expertise.

VII. Coordination of Supply and Demand

Reference has already been made to the existence of fluctuations in demand, over extended periods of time, in terms of quantities and differing species. The analysis of such trends is of value to the commercial breeder and to a lesser extent the user-breeder. We are concerned in this section with the more troublesome problem of short-term fluctuations.

The period of useful value for experimental purposes is short in most species. Where small rodents are used in experimental assays this period may be as short as 10 days, since age and weight requirements are set within narrowly defined limits. Very large numbers of animals are used within this experimental category, and if demand is not very consistent, considerable wastage may result. In an attempt to remedy the situation, reduced breeding may result in an insufficient number being available in the defined range, again resulting in wastage and, additionally, delays to research programmes.

Where larger animals are bred, especially cats and dogs, the output tends to be seasonal and fluctuations may also result from too small a number of breeding females. If the demands tend to fluctuate or the requirements are for groups of animals, this, together with breeding fluctuations, may result in the necessity to hold large numbers until they are used. The cost of maintaining these species is large in terms of labour and supplies.

Wastage and the necessity to hold animals against possible future use is an extravagance of time, effort and expense. Two ways of reducing this waste are open—by cooperation with other laboratories and by careful coordination of supply and demand.

A. COOPERATION WITH OTHER USERS

Liaison with other users can do much to iron out the peaks and valleys of production and usage. This can be particularly valuable in the case of dogs, cats, rabbits and guinea-pigs. Most users of these species obtain supplies from external sources and the user-breeder should find little difficulty in disposing of surplus stock. Even when sales have to made at prices below cost, the savings on maintenance usually justify disposal. The movement of surplus small rodents may introduce problems associated with the introduction of disease or interpretation of experimental results due to strain differences. The user-breeder embarks on his policy because of these considerations and others discussed in Chapter 9 and will introduce animals from outside only as a last resort. However, where adequate facilities for isolation exist and when strain considerations are not important, mutual arrangements for the use of surplus stock may be financially advantageous.

Where a number of laboratories are situated within reasonable distance of one another, cooperation can be arranged without difficulty. In addition, and of particular advantage to the more isolated laboratories, nationally distributed information relating to surplus stock is available in the *Parade State* published every 2 weeks by the Laboratory Animals Centre in Britain. This facility is now available in many countries; in the United States the *Continuing Inventory Program*, published bi-weekly by the United States Institute of Laboratory Animals Resources, is an example. Such services deserve wider publicity and extension in order to reduce wastage of animals. The limitations are mainly related to the distances animals have to be transported, but they do serve to reduce the possibility of surplus culling in one laboratory where another nearby is suffering from a shortage.

Exchange of information of this kind is undoubtedly useful in respect of animals of fairly general specification, but it should not be overlooked that it can be of still greater value in the case of animals of special strain or of species not commonly used or available commercially. The essential requirements remain that such lists are up to date, published frequently (the minimum useful interval is 2 weeks) and widely received and contributed to.

B. INTERNAL COORDINATION

The director of a research programme who outlines his animal requirements once a year and then considers that his responsibilities have been discharged will be plagued with disappointments and frustration. The very nature of biological experimentation, with the possible exception of routine assay work, produces experimental change and revision of plans at frequent intervals. It is essential that those with administrative responsibility for supplying and maintaining experimental animals are kept fully informed of such changes—be they only possibilities or firm proposals. The main responsibility for imparting such information lies with the experimenter, who only too frequently waits for someone on the animal side to extract the information from him. The breeding situation and availability of accommodation must be under constant review and this is best accomplished if placed on a formal footing. In an organization where a number of separate experimental programmes using a variety of species are undertaken, intervals of 3 months are not too frequent for such meetings. They should be attended by the senior staff on both the animal and experimental sides so that the full implications can be discussed and a policy agreed which carries the full weight of responsibility. The cost of supplying and maintaining animals will be discussed in the next chapter and will be seen to constitute a large proportion of the cost of experimentation. When these considerations are taken into account, together with the financial implications of delayed research, it can be appreciated that internal coordination of supply and demand is to be taken seriously.

The basis of such meetings is the presentation of detailed and comprehensive statistical data relating to past usage, present availability and breeding levels, and future planned usage. Such data can be assembled from weekly breeding and usage figures maintained by unit heads and should be presented in such a way that proportional wastage is apparent and the relationship between planned usage and availability is clarified. On a more informal basis, monthly meetings between senior animal and laboratory technical staff can do much to solve the temporary supply problems and provides a valuable forum where feed-back information can be provided by laboratory staff concerning problems arising during experiments and clinical and behavioural observations.

The user-breeder of dogs and cats and, to a lesser extent, of rabbits and guinea-pigs faces a problem of long-term forecasting of usage of these species. Since the time period required for the increase of breeding colony size ranges from several months to 2 years, considerations of colony size can only be made at intervals of about 1 year. Overproduction is not the answer to the problem even if a ready market exists, since the cost of user-breeder production is invariably greater than the commercially available animal and the cost of maintaining such animals against future usage is very high. The only way out of the difficulty lies in very close cooperation between the experimeter and the chief animal technician. For any cooperative system to work there must be compiled, at regular intervals, a chart showing the numbers of young animals in various age groups. This information, compiled from a growth-rate graph, will indicate how many animals may be available for experiment during the growth period—4 to 6 months for cats or 6 to 9 months for dogs. These data, together with forecasts of usage over that period and an assessment of seasonal breeding effects for the period following, will enable decisions to be made on the advisability to retain young stock or dispose of a surplus.

VIII. Hazards

The experimentation conducted by a laboratory may involve the use of animals as carriers of substances or organisms potentially dangerous to man. Under these circumstances special techniques in husbandry must be carried out and it may even be necessary to employ equipment specially designed for the purpose. In Chapter 6 mention has been made of such equipment and modifications to animal house design described, with a view to the elimination of danger to staff working with such animals.

Two main hazard groups can be considered. First, the use of radioactive isotopes in experimental animals, and second, the conduct of research into human pathogens using animals as carriers. Although the disciplines to be employed in husbandry, handling and disposal are similar to some extent, sufficient differences and special considerations exist to merit discussion under separate headings.

There is also a number of miscellaneous hazards which are not necessarily peculiar to animal houses. A useful symposium on animal house hazards has been published by the Laboratory Animals Centre (1961).

A. RADIOACTIVE MATERIALS

We are concerned here with the use of radioactive materials in tracer experiments, which form the bulk of experimentation with these materials. Where large quantities of highly radioactive substances are used, the considerations are mainly concerned with animal house design and special staff clothing and disposal methods. These have already been dealt with in Chapter 6.

In principle, the housing of contaminated animals should be confined to a room reserved for this purpose. The cages should be distinctly labelled and the date and quantity of administration of the isotope clearly marked. Instruments and cleaning equipment must be kept separate from uncontaminated animals and the remainder of the animal house. Staff movements must be restricted to the minimum required for normal husbandry operations and experimental routines. Special care must be taken to avoid wounds inflicted by the animals or direct contact with excreta; ventilation should be checked frequently to ensure adequate removal and dilution of exhaled air and aerosol contamination by animal coughing and movement. Where a possibility exists of contamination spread by vermin, such a room should not be used.

Decontamination procedures depend on severity but for the most part are comparatively simple to carry out. A change of outer clothing should always be carried out by animal house staff on entering or leaving the contaminated room; hands should be washed and scrubbed with a soft brush using a plentiful supply of water. Monitoring after washing is always advisable.

The decontamination of equipment depends on its life and cost. Where possible, disposable instruments should be used.

The decontamination procedure should be undertaken immediately and usually can be accomplished by copious washing, using a brush to dislodge material. This procedure should be used for cleaning cages and racking; again, subsequent monitoring is advisable before returning cages to the clean store. Where the half life of the radioactive contaminant is short, it may prove more convenient to store contaminated equipment in a room set aside for the purpose, until the level of activity has reached an acceptable value.

Radioactive waste includes bedding, tray papers, unused food and cadavers. These should be confined in heavy-duty plastic bags and labelled accordingly, unless disposal is immediate. Cadavers which are not disposed of immediately should be stored in a deep-freeze or sealed in plastic bags to prevent the spread of contamination by the process of decomposition. The disposal of radioactive waste can be carried out by release into drains and sewers, incineration, release as a gas or aerosol into the atmosphere or burial. The usual method of disposal of material arising from an animal house, described above, consists of incineration using suitable equipment. Whatever system is used, consultation with the radiological health and safety officer of the local authority is essential.

The responsibility for the use and disposal of radioactive materials rests with the responsible and competent persons within the laboratory in consultation with the local authority, and such work must not be undertaken without the necessary official approval and licensing. Detailed and authoritative information can be obtained from the Radioactive Substances Act (Parliament, 1960), the publication by the International Atomic Energy Agency, *Safe Handling of Radioisotopes* (Report, 1962), *Radiological Protection in Universities* (Report, 1966), published by the Vice-Chancellors' Committee of the Association of Universities of the British Commonwealth, and the Department of Employment and Productivity booklet *Codes of Practice* related to radiation hazards (Dept. of E. and P., 1968).

B. INFECTIVE HAZARDS

Infective hazards in the animal house are of two kinds. There are the infections that the animals may carry, because no special steps have been taken to eliminate them. These will include infections brought in with animals captured from the wild, such as B virus in macaque monkeys; accidental infections carried over from a breeding colony, such as lymphocytic choriomeningitis in mice; and casual infections that may become established in the animals, doing more or less harm to them but dangerous to man, such as leptospirosis in rats or salmonellosis in many species. There are many more infections that are in every way undesirable and should be eliminated or excluded.

But in some experimental work it is necessary to work with infection deliberately introduced, which can carry serious dangers for man. For example, guinea-pigs used for the diagnosis of tuberculosis develop a more or less generalized infection in positive cases, and will excrete live tubercle bacilli. Monkeys used to test poliovaccine can transmit poliovirus. Any potentially dangerous pathogen, in the investigation of which animals are used, presents a hazard to those working with it, and the hazard is increased if the infection is established in experimental animals. For the hazards of handling monkeys, reference should be made to *Laboratory Animal Handbook 4* (see Bibliography).

In relation to both kinds of hazard it is necessary to take proper precautions. These may include elimination, isolation, protection, separation and therapy.

1. Elimination

As far as possible, incoming animals should be checked for the presence of human pathogens, and if any are found the animals should either be destroyed, or the infection cleared up by suitable treatment. The design of an animal house that may take in, for example, wild caught monkeys should make provision for keeping new animals in strict quarantine during the period of examination and elimination. The reception or quarantine area may well be a completely separate establishment.

2. Isolation

In some case it may not be possible to eliminate all chance of the presence of a dangerous infection, and the animals will then have to remain more or less permanently in isolation, even when the time comes for them to be used experimentally. The animal house will then have to possess a secure barrier to ensure that no human pathogen can escape from the colony or infect those who work with the animals in it. The same sort of precautions will be needed in cases where work is being done on diseases of economic importance in animals; for example, foot and mouth disease in cattle, where it is vital that no accidental escape of the virus can occur through the medium of the staff working with it.

3. Protection

Those who work with poliovirus are always protected by vaccination. The same precaution is normally taken with all staff where work may bring them into contact with human pathogens; with the tubercle bacillus (BCG vaccination), or any other dangerous microorganism, against which protective inoculations are available. Not to provide adequate protection for all staff at risk would probably be regarded in a court of law as culpable negligence.

4. Separation

In nearly all cases of experimental work entailing the infection of animals with dangerous pathogens, it will be possible to separate the infected animal and its immediate surroundings from the staff working with it. Infected animals may be kept in special cages, through which is drawn a constant stream of air which is burnt or otherwise rendered safe after it has passed through the cage. In some cases a whole room may be treated in this way; those entering will wear protective clothing, including at least a respirator and perhaps also something resembling a space suit. In other cases, the use of special overalls, gloves and masks will be considered sufficient. In all cases there is a physical barrier separating the source of the infection, the animal, from the persons working with it. Once again, provision will need to be made for the exercise of these precautions, in the design of the animal quarters.

5. Therapy

Despite all the foregoing precautions there is always a chance that infection may get through to the staff or to healthy animals elsewhere in the animal house. If this happens, there must be a proper therapeutic programme. The staff member's own doctor must always be informed of any special hazard to which his patient may have been exposed. If there are any specific therapeutic measures that may be taken, these should always be available for use on medical advice. In the case of spread of infection to other animals, whose consequent destruction cannot be contemplated, specific therapy may have to be given to them

C. OTHER HAZARDS

There are many miscellaneous hazards in the animal house that are not necessarily peculiar to it. In any room where a cement floor is washed down from time to time, there is a risk from unsuitable or damaged electrical equipment. The use of certain disinfectants, which leave a residual film, can make the floor dangerously slippery; one amphoteric disinfectant is particularly hazardous in this respect. Poor ventilation, or an abnormal accumulation of dust, can give rise in time to specific sensitivities, even allergies, in susceptible individuals. Some disinfectants or other substances used in the animal house can give rise to skin lesions. Animals sometimes bite, and these bites may be dangerous, not only by way of direct injury, but also because of resultant infection.

Injuries acquired from damaged equipment are not infrequent. One particularly disagreeable accident can be caused by the broken shaft of a glass drinking tube; a cut index finger, even a cut flexor tendon, which is a crippling injury, can result from this.

In short, accidents happen, even in a well-conducted animal house. When the animal house is a rigidly barriered one, with inadequate visual and auditory communication with the outside, an accident to a worker on his own may not be noticed at the time, and the victim may be unable to call for help. It should be a rule, therefore, that in such animal houses suitable steps are taken to deal with this special hazard. It is safest if no one ever enters by himself; but if it is ever necessary for someone to do so, then there must be a routine for keeping in regular touch with someone outside, during the whole period. There is no record so far of a member of staff being found dead and cold within a barriered animal house, but unless proper precautions are taken it is only a matter of time before such a tragedy occurs.

CHAPTER 9

Economics

I. Costs in Relation to Research

Biological research relating to the functions and behaviour of animal species requires the provision of whole animals or parts of the animal body. This truism can be readily acceptable, but the need for an animal system functioning in a normal manner is often overlooked. In other parts of this book the requirement and desirability of "healthy" animals is discussed at length, but in considering the economics of providing laboratory animals, the effects of using animals which are "unhealthy" or not maintained in a suitable environment must be taken into account. It is not unusual to meet situations where long-term toxicity experiments have to be repeated because of death or disease arising in the test animals. Similarly, complex testing in large animal species, involving preparation of apparatus and pre-treatment over a period of days, may fail due to mortality of the test species under conditions of general anaesthesia. Assay systems developed for the testing of potential therapeutic agents may produce results which are not comparable over a period of time, due to variations in the test animal. These situations can be very costly and in many instances can be avoided by proper consideration of the animal requirements before experimentation is commenced and in planning animal accommodation and breeding pro-grammes. The cost of such occurrences can be demonstrated using two examples.

A 3-month toxicity test in 200 rats may cost £400 in labour, overheads, maintenance of animals and the provision of the test compound. This, however, is only the apparent cost. To this must be added, if the experiment has to be abandoned, for a commercial organization, the loss of 3 months' sales and the possibility of a competitor getting on the market first. Also, the frustration produced by such an event, on senior and junior staff, is damaging to their efficiency.

A biological research group involved in fundamental research, consisting of four graduates and eight supporting staff, may cost in the region of £175 per day including overheads and supporting animal facilities. The failure of a cat under anaesthetic to last the period of an experiment of several hours' duration, due to chronic respiratory disease, will involve the cost of that day's work. If other research groups are dependent on the result of that experiment for the formulation of their own plans, their costs can be added. When consideration is given to the fact that, in all probability, the difference in cost between the purchased animal which failed, and one available at higher cost in a satisfactory condition, was only a matter of a few pounds, the false economy of not relating animal quality with the purpose for which it is required becomes apparent.

On the other hand, it is not necessary to use animals of a high quality for many purposes—the provision of organs and tissues for *in vitro* experi-mentation for instance.

The cost of providing laboratory animals is frequently a source of friction between the research workers and the management executive or group responsible for providing such funds. It is hoped that this discourse helps to emphasize the false economies inherent in the attitude that the provision of laboratory animals is the poor relation in the economy of biological research. In the course of this chapter, as in Chapter 6, it is intended to emphasize also that lavish expenditure does not necessarily provide im-proved animal quality.

Later in this chapter, consideration will be given to the compilation of operating budgets for the provision of animals and animal facilities. At this point it is worth while considering the preparation of overall research costs related to animal costs. For this purpose, three categories of biological research are considered:

1. Toxicology and experimental pathology.
    High cost of animals and animal maintenance in relation to de-partmental costs.
2. Physiology and pharmacology.
    Small numbers of high-cost animals but large numbers of low-

cost species. Maintenance costs low since most experimentation is
acute.
3. Biochemistry.
    Large numbers of low-cost animals used; pre-treatment may raise
    maintenance time.

Approximate cost proportions for these examples are given in Fig. 9.1 on
the assumption that a balanced activity exists within the experimental unit.
The variations between these selected examples demonstrate the need for
methodical planning on both short- and long-term bases, if research funds

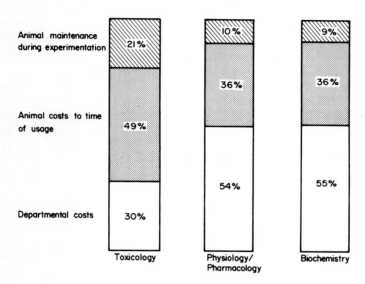

Fig. 9.1. Animal costs related to departmental costs.

are to be utilized to the maximum advantage. When such proportional
costs are considered in relation to the provision of supplies for experimental
units not requiring laboratory animals, it can be appreciated that changes in
direction or intensity in biological research can have financial implications
which are often overlooked.

II. Operating Budgets

A. INTRODUCTION

A meaningful operating budget is compiled on the basis of forecast data
provided by the users of animals and facilities. This is frequently the most
difficult part. Where a new activity is being set up it can become a matter of

the intelligent use of crystal balls. A budget compiled on the basis of "more or less what we spent last year" may save time, but is valueless for the purposes of cost control (about which more later) and the forecasting of cost effects resulting from changes in research direction. The main requirements for the intelligent compilation of an operating budget are therefore (1) thoughtful forecasting of usage, and (2) breakdown of such forecasts into the component parts of a budget.

B. FORECASTING

This activity must be undertaken by all research personnel responsible for carrying out a research programme based on the use of animals. The proportion of animal costs to the overall departmental cost has been shown to be high. The forecasting activity must therefore be undertaken with this fact in mind, since bad forecasting can result in actual expenditure differing widely from the calculated budget. In research units where the approved budget becomes a fixed allowable sum, research can be hampered by insufficient funds or unnecessary spending undertaken in order to achieve budget. The temptation to overbudget to allow for contingencies is real and costly. The ability to forecast to within 10 per cent of the actual usage is also extremely difficult in most cases. How then can a usage forecast of value for budget purposes be compiled? The nature of biological research is such that changes in direction may be rapid, but a forecast should be possible for an expected usage over a period of 12 months. Having arrived at the expected usage, an attempt to analyse the probability of varying from this figure by steps of percentage variation should be made. It may be possible that at a particular usage level, additional facilities may be required. Figure 9.2 illustrates a graphical means of presenting such data. By combining such data from a number of departments a reasonable forecast of usage can be built up. The degree of contingency can then be applied to the

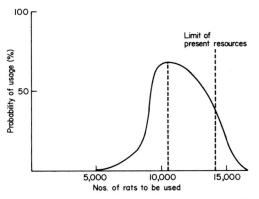

Fig. 9.2. Forecast usage of rats over 12-month period.

final stage as a result of decisions made at professorial or divisional management level. Decisions may also be made on the provision of new facilities or reallocation of existing space.

The degree of forecasting accuracy has different implications for different species. Obviously, the cost effects of variations from the expected are greater for large animals than for the more easily bred laboratory rodents. Where the larger species, such as cats and dogs, are not purchased but bred within the organization, accuracy of forecasting is essential. Output levels for these species take many months to be adjusted and the provision of new breeding stock within a closed colony reduces the number available for experiment.

Once a decision has been reached on the forecast level of usage, this must be converted into the practical terms of breeding output or purchase. A level of wastage is inevitable unless production, within the facilities available, cannot meet demand and a minor proportion of the requirement is purchased. Particularly in the case of rodents used in substantial quantities within closely defined age and weight limits, wastage of animals must occur. This must be brought into the calculations before arriving at a production or purchase level. Where larger animals, such as cats and dogs, are bred within an organization but usage is sporadic, considerable cost implications will accrue due to the high cost of maintaining such animals and the inflexibility of varying the breeding levels. While every effort has to be made to meet the stated requirements of the experimenter, a heavy responsibility rests with the latter in ensuring that his defined requirements in terms of quantity and age and weight limits are realistic in the context of his experiments. The implications of unnecessary wastage are ethical as well as financial.

C. COMPILATION OF THE OPERATING BUDGET

Although an approved budget constitutes a single sum for the provision of an activity, it is axiomatic to state that it consists of a number of individual parts. At the start of the process of compilation, a decision must be made at what level a group of items will be considered separately. For example, will an estimate be made of the cost of paper towels in a unit, or will these be included in a general heading such as "domestic expendables"? A useful guide is to consider individually all items representing over 1 per cent of the total budget. For instance, in a cat-breeding unit costing £12,000 per annum to operate, if paper towels cost £150 per annum these should be itemized. While the bulk of the costs will be taken up by such items as food and wages, the identification of smaller items is useful for accurate budget compilation, forecasting cost implications of expansion decisions and expenditure control.

On the broader aspect, where animal facilities comprise separate units for breeding and maintenance of different species, each unit should be considered separately. Again, the advantage of this approach is that annual variations in demand can be interpreted more accurately for financial purposes and more realistic estimates relating to proposed changes in facilities can be made more quickly.

Examples of cost breakdown for budget purposes are given in Table 9.1. The proportions are only approximate and are intended to illustrate the variations which can be expected where the functions are dissimilar. The actual proportions will vary according to the quality of animal required, the type of labour available, the situation of the laboratory and the nature of the research undertaken. These factors are discussed in greater detail in Chapter 6.

Table 9.1

Variations in cost breakdown for differing types of animal facility

	Dog breeding unit, per cent	Small rodent breeding unit, per cent	Experimental unit, per cent
Salaries and wages	33	40	20
Supplies	16	16	24
Overheads			
Services	18	21	12
Maintenance	11	13	6
Depreciation	22	10	38
Totals	100	100	100

The basic items for consideration when compiling an operating budget can be listed as follows:

Salaries and wages
    Basic (plus bonus where applicable)
    Overtime
    Casual labour
Food
    Basic diet (itemized where applicable)
    Supplements
    Special diets
Bedding
Purchase of animals
Miscellaneous
    Protective clothing

Laundry
Domestic supplies (soap, towels, tea, coffee, sugar, etc.)
Disinfectants
Laboratory supplies
Veterinary supplies (antibiotics, vaccines, etc.)
Office supplies
Disposable containers (food, water, soiling trays)
Sacks
Refuse disposal
Capital equipment
Repairs
Replacements
New items (equipment written off over a period of years should be included under overheads—depreciation)
Transport
Fuel and oil
Tax and insurance
Maintenance and repairs
Personnel expenditure
Travel costs
Attendance at courses and conferences
Personal telephones
Overheads
Services
Electricity, gas, water, oil
Maintenance
Buildings
Contract servicing
Site maintenance
Security
Charges from other areas
Financial
Secretarial
Personnel
Depreciation
Buildings
Equipment

Where a number of units form part of a large animal facility, each unit should be charged with a proportion of the costs of general services such as transport, managerial and secretarial.

This list can be used as a check list and guide. Experience in the operation of a particular unit will determine which sections can be reduced and which expanded.

A useful final layout is shown below for an imaginary facility comprising breeding and animal maintenance units, where the headings previously described are compressed into main groups. This layout will provide data at a level which is more readily appreciated by management executives responsible for the approval of funds.

Table 9.2

Breeding units

	Dogs	Cats	Small rodents	Rabbits	Maintenance experimental	Totals
Salaries and wages						
Supplies						
Overheads						
Animal purchases						
General services						
Totals						
Less sales						
Grand totals						

Such documentation produces a cost awareness in those responsible for deciding the direction of research programmes.

III. Cost Control

The broader concepts of cost control involve considerations at the design stages for new facilities or modifications to those existing, decisions regarding breeding techniques, the quality of animals to be used and staffing. These subjects are discussed elsewhere in this book. It should, however, be stressed that once a decision is made it cannot be regarded as permanent. New techniques and equipment are constantly being developed and these should be reviewed in the light of each specific operation. The interchange of ideas and data between laboratories, through attendance at conferences and visits to other facilities, is not superseded by reading books or by maintaining an unshakeable faith in the absolute virtue of one's own approach. Of all technical disciplines the science of laboratory animal management is most subject to the specific requirements and conditions of individual laboratories. The most satisfactory operational method for any particular unit must be the result of an intelligent distillation of available knowledge. The achievement of this ideal at the lowest cost should be the aim of all who are concerned with the management of laboratory animals.

This principle applies, in the final analysis, to all staff. Cost consciousness is not the prerogative of the higher echelons, but should be the concern of all who use apparatus, machinery and expendable stores. Cost control is also the utilization of staff to the best advantage; the technician who suggests a method for doing the same operation in a shorter period of time is not only improving unit efficiency, but also is satisfying his own desire to work as an intelligent member of a team whose purpose is not just to perform a routine task but to improve and innovate.

The key member of a laboratory animal division team, where cost control is concerned, is the head of a unit of defined function—experimental dog unit, rat breeding unit, and so forth—who can be referred to as the unit head. He is in constant touch with the day to day operation of the unit, he is familiar with the minutiae of the techniques employed and he is involved with decisions relating to the organization of the unit to fulfil the requirements of the animal users. One of the prime responsibilities of the unit head is to seek out methods whereby efficiency is improved. It is his responsibility to translate forecast requirements into the practical terms of an operating budget. It is also his responsibility to ensure that breeding techniques, early culling and accurate records provide the minimum of wastage. Having compiled the operating budget, it is the unit head's responsibility to ensure that the target figures for the individual items comprising the budget are not exceeded without reasonable justification. Cost control in this context is extremely difficult, if not impossible, without a detailed breakdown of the budget as previously described.

The method whereby expenditure is recorded must depend upon the accounting system used within the organization. Whatever the system, a ledger should be maintained by the unit head and reviewed on a monthly basis. Such a ledger should be based on the same breakdown as the operating budget. Accounting systems based on computerized recording of invoice statements lend themselves particularly well to this technique, since items are easily identified and can be entered into the ledger on a monthly basis. Systems which are not based on a computer involve more time, but recording can be made at the time of invoice authorization. If recording is undertaken at the time of placing an order, errors may occur due to new prices, hidden costs such as transport or duty, delays in delivery or subsequent modifications to or cancellation of the original order.

Experience has shown that such a system is not difficult to master, even by the most financially disorientated of technical supervisors, and involves only a small proportion of the unit head's time. The advantages that accrue are twofold. An accurate and detailed cost analysis becomes available, from

which trends can be determined, projections can be made and disproportionate expenditure can be identified. The second advantage is psychological. Targets can be set and all staff become involved in the efficiency of
their unit—from a financial as well as a technical standpoint. To show that
it is possible to breed a healthy experimental animal is a satisfying technical
achievement; to produce that animal at a lower cost than the previous year,
in spite of external rising costs, produces a sense of satisfaction with a
beneficial effect on morale. Experience has shown that all levels of staff can
respond positively to cost awareness, when this is properly presented—
and they enjoy it.

IV. Investment of Capital

Many surveys have been undertaken in the laboratory animal field, on
technical and administrative matters; there is one outstanding topic which
has been neglected. While a few units have provided proportional cost
data (Walker and Stevenson, 1967), the reporting of absolute costs and
relationships within overall departmental and divisional expenditure is con-
spicuously absent. The reasons for this may be twofold. Too often the
provision and maintenance of the animals used in biological research are
regarded as the poor relation and no attempts are made to evaluate this
function in any depth. It is one of the aims of this book, and this chapter in
particular, to stress the need for such education and to place the laboratory
animal in its proper context; the failure of costly research due to neglect of
these considerations has previously been illustrated. Secondly, it is suggested that when the real cost of the animal operation is examined in depth,
many laboratories take fright and psychologically look the other way. The
planning of a programme concerned with biological research must contain
due consideration of the animal requirements—in depth—and decisions
made in the light of comprehensive information. The tendency to make
decisions on research direction and depth, based on laboratory considerations only, often results in frustration and waste of expenditure due to the
lack of capacity or funds to provide the required experimental animals.

The proportion of animal provision to overall departmental costs has
been discussed earlier in this chapter. Again, it should be stressed that these
figures are approximations; nevertheless, the proportions are substantial
and the considerations given to animal facilities should be as detailed and
comprehensive as for laboratory facilities where investment of capital is
envisaged. While much of the equipment for a new animal house is readily
available and does not suffer from extensive delivery delays, as encountered
with many of the more sophisticated items of laboratory equipment and
apparatus, the details of animal house furnishings and services may be
comparatively more complex. Completion times for these buildings may
extend for months beyond the estimated dates due to the factors already

discussed in Chapter 6. Laboratories can often function satisfactorily during the post-completion period, when minor adjustments, alterations or modifications are being made. It is often impossible to move animals into a new facility until after such operations have been carried out. It is apposite to point out, in a chapter devoted to economics, that considerable costs may accumulate due to a time lag between the practical completion of laboratory and animal facilities.

A recurrent theme found in Chapter 6 related to the fallacy of supposing that excellence was a direct function of investment. No apology is made in stressing this again. While under-investment of capital or budgeting for operational costs below a certain point can make nonsense of the financing of a research programme, to assume that lavish expenditure will cure all problems is ultimately fallacious. Before capital is invested the following decisions must be arrived at:

1. The quality of the animals required and the acceptable limits of risk.
2. The maximum quantities to be bred or housed for each species.
3. The degree of space flexibility required.

These basic requirements can then be translated into practical terms by those technically qualified to do so. At this stage there is no better guide than discussions with other laboratories. The pitfalls are many and each one can be economically expensive.

Finally, it should be remembered that the construction of modern laboratories can be undertaken to provide considerable flexibility—both in relation to alteration of special facilities and complete change of function. This is usually not possible with animal house construction, either as separate buildings or as part of an existing structure. If a wrong decision is made at the planning stage it may prove extremely costly to rectify. Also, short-term thinking with resultant inflexibility will produce inadequacies in the animal support facilities as the research programmes change in direction and emphasis.

V. Relative Costs of Species

The use of a particular test species for the elucidation of a biological problem is usually made on consideration of the experimental techniques to be used, numbers of animals required, physiological species differences and facilities available. The increasing sophistication of modern response-monitoring equipment has tended to promote the use of fewer animals but of larger species. For example, the screening of potential drugs, undertaken by the pharmaceutical industry, has tended to move away from recording large numbers of responses in mice in comparatively primitive tests to fewer, more definitive, responses in rats. In many cases a swings-and-roundabouts effect has resulted in no increase in animal costs. However, where research direction has resulted in a swing towards the larger animals, such as rabbits, cats and dogs, a considerable escalation of costs has resulted, in spite of the reduction in numbers of animals used.

In considering relative species costs, generalizations must be made, since actual costs depend on the facilities available. In broad terms the cost of producing different commonly used laboratory animals by the user are given in Table 9.3.

Table 9.3

	Approximate annual production	Operating cost	Overheads	Total
Mice				
to age 5 weeks	10,000	6p	5p	11p
Rats				
to age 5 weeks	20,000	25p	20p	45p
Guinea-pigs				
to age 3 months	1,000	£1·80	£0·80	£2·60
Cats				
to age 6 months	600	£24·50	£4·75	£29·25
Dogs				
to age 6 months	200	£35·00	£24·00	£69·00

These examples are taken from a unit where capital outlay on buildings was small, as a result of using timber as the main constructional material, in the case of rats, mice and cats. In spite of shorter write-off periods (5 years) applicable to temporary buildings, the annual depreciation was low. In the case of dog breeding the planning intention was mainly for experimental purposes and the construction was therefore carried out as a permanent building; the longer write-off period in this instance (30 years) did little to alleviate a heavy depreciation factor in calculating costs per animal bred. The advantages of using cheaper building materials have already been discussed in Chapter 6. In the illustration given here the use of wooden buildings for dog breeding would have saved about £10 per dog in overheads. The breeding of cats and dogs presents a similar proportional pattern in terms of the cost of labour, food and services. The advantages of using timber buildings for cat breeding are shown in this example by the relatively small overhead proportion. This is also demonstrated by the high overhead proportion in guinea-pig breeding costs, which was undertaken in an animal house situated within a modern laboratory block.

While it can be seen that the cost of user production of rats and mice, when undertaken at a reasonable output level, is not greatly in excess of the cost of purchased animals, a considerable discrepancy is apparent in the case of cats and dogs. When user-producers of these species discuss costs it is not unusual to hear overall figures for cats of £40 per animal and £100–150 for dogs. One breeder of dogs in a continental country estimates the cost to be £250 per 6-month-old dog.

The figures for small rodents relate to strains which are random bred within a closed colony. These strains produce average litter survival numbers of 9–12 for rats and 12–14 for mice. For the user-breeder of inbred strains of mice and rats the cost picture may be very different. The authors have both experienced the depressing effects of breeding certain inbred mouse strains, where an average litter size of four produced every 2 months is hailed as a successful achievement.

The choice of experimental species also produces repercussions on holding facilities. As an illustration, a 6 m × 3 m experimental holding room will accommodate the following:

Mice	4800
Rats	840
Guinea pigs	300
Rabbits	75
Cats	40
Dogs—short term	6
Dogs—long term, where exercise facilities are included in the 18 m²	2

These numbers are based on the assumption that adequate environmental provisions are available. It is sufficient to point out that any change in holding requirements on the part of experimenters should be discussed with animal house supervisors as early as possible and the cost effects of such changes should be borne in mind.

VI. Procurement

A recurrent question facing research directors and individual experimenters is: "Do we buy from outside the organization or breed our own requirements?" There is never an easy answer to this question, but basically it is a question of economics. In practice, a decision may be reached after careful and extensive consideration of all the factors involved, only to be frustrated by a sudden requirement for a new species or rapid increase in demand for one already bred, which cannot be satisfied within the existing facilities. Nevertheless, the decision has to be made in the light of prevailing conditions and foreseeable trends.

Purchased animals are not cheap animals. Unless an economic price is paid, the commercial producer will go out of business or an unsatisfactory animal will be supplied. However, unless very large numbers of a particular species are to be used it can be assumed that user-breeding is more costly than external purchase. With the larger experimental animals even this does not apply. What then are the advantages of the two methods of procurement? The following lists can be used as a guide when considering this problem.

Advantages of user-breeding:

1. Animals are fully under the control of the user, consequently he can be confident of information regarding:
   a. genetic history;
   b. administration of non-dietary substances;
   c. incidence of disease;
   d. incidence of observable, non-traumatic peculiarities such as localized alopecia and chromodachryorrhea;
   e. growth-rate statistics.
2. Detailed background biochemical data can be obtained.
3. By periodic microbiological monitoring the disease state of the colony can be kept under review.
4. The user is more confident of continuity of supply and uniformity of strain.
5. Strains can be selected and characteristics modified, according to the user's requirements.
6. Experimentation concerned with reproduction can be undertaken under conditions where extensive control data are available.
7. Healthier animals of certain species can be produced.
8. Adverse stress effects resulting from transport can be avoided.
9. Higher calibre animal staff can be recruited and kept where there exists the additional interest of animal breeding.
10. For long-term toxicological experiments in rats, slow-growing strains which do not tend to produce an excess of fat are advantageous. The commercially obtainable rat tends to have fast growth characteristics for economic reasons.

On the negative side there are psychological factors which may be real or imaginary; fear of the commercial supplier changing strains or even going out of business, distrust, sudden cessation of supply due to disease outbreak, effect of unknown stress factors due to transportation or environmental change. In some cases these considerations may be relevant; in others, because of the requirements of the research undertaken, they may be quite unimportant. Objectivity may be difficult to achieve since these factors may operate at the subconscious level.

Advantages of external purchase:

1. The initiation and management of a breeding programme, within an experimental animal facility, requires a higher level of expertise than for a holding operation alone. This is based on the assumption that

high-quality animals are required or that larger species than the small rodents are to be bred. The manpower and cost implications for management consideration have therefore to be supported by pressing reasons why animals should not be purchased externally.

2. Small rodents can be obtained commercially of very high quality and a wide range of strains are available. Gnotobiotic and germfree animals are also available.

3. Wide fluctuations in demand are more economically met by external purchase on requirement.

4. User-breeding of larger animals takes many years to establish and colony size adjustments take many months to stabilize. Overproduction of these animals can prove extremely costly since holding costs are high. Even where usage figures taken annually tend to be reasonably constant, if the monthly demands exhibit wide fluctuations, holding costs against expected demand can greatly inflate the cost of the animal at the time of use.

5. The control of the animal supply and holding function is often laid on the shoulders of a research scientist. By relying on a commercial supply of animals to meet the requirements of a research group, the scientist can devote more time to his own experimentation.

6. Animals may be purchased in a post-operative state having been subjected to, for example, thyroidectomy or hypophysectomy. This relieves the user of training staff, obtaining Home Office licences for such procedures and delays to research. Because such operations are performed in large numbers by very experienced staff employed by commercial suppliers, they are probably more efficiently undertaken than by the user whose experience is more spasmodic.

In considering these factors the degree of importance to be attached to any one aspect depends on specific circumstances related to the nature of the research undertaken, the facilities available and the funds allocated. In this latter respect it is again relevant to comment that research frustrated through the use of unsatisfactory animals results in economic waste out of all proportion to the animal costs.

CHAPTER 10

Production Techniques

I. Introduction

Keeping a few pairs of animals, or a colony of a few dozen breeders, calls for no great degree of organization or expert knowledge of breeding techniques. Maintaining a colony of several thousands of breeders is quite a different problem. The contrast may be compared with growing a row of cabbages or peas in the back garden, and running a market garden of several hundred acres.

Rats and mice stand out from all other laboratory mammals in the numbers in which they are bred, and the relative concentration of large-scale breeding in a small number of breeding units. Although guinea-pigs may be used in numbers approaching those of rats, both their use and their production tend to be more scattered. This chapter will have particular reference to rats and mice, mainly because they are the laboratory species in which sophisticated large-scale production techniques have been developed. But much of the content of this chapter could apply equally to other species that were required to be bred in large numbers. Comparisons with the poultry industry will suggest themselves from time to time.

In Chapter 1 it was seen that mice account for perhaps 70 per cent of all laboratory mammals, and that rats may make up for a further 15 per cent. There are very many strains of mice, including 200 or more inbred strains as well as a considerable number of outbred. The laboratory demand is mainly for outbred mice, but cancer workers and geneticists call for inbred strains, and for certain types of work there is a need for first generation (F_1) crosses between inbred strains. Occasionally F_2 crosses, or mosaic populations (see p. 49), are needed.

There are also many inbred strains of rats, but they are far fewer than those of mice. Less frequently than in mice there is a need for F_1 crosses also.

II. Primary Type Colonies

All laboratory animals are bred in colonies, more or less closed to the introduction of outside blood, and some of them have remained strictly closed for so many generations that the animals have become quite distinctive in appearance and other characteristics, even though within the colony closely related mating is avoided and the colony is large enough to ensure a low coefficient of inbreeding.

In the case of inbred strains the colony remains closed by definition, and the rules for maintaining and designating inbred strains of mice are strict and practical (Staats, 1968). But whether we are dealing with inbred strains or outbred animals, colonies tend to develop their own characteristics, which may by selection or other means have specific application. There will be a demand for such animals, and breeding units will tend to multiply and in doing so perhaps lose some of the specific characteristics of the original colony. It is therefore desirable to recognize in some way the original colony, so that others can be referred back to it if necessary.

Recognition is in the name "primary type colony": primary, because it is the first link in a chain of production of experimental animals; and type, because it conforms to a specification that is relevant to the needs of the user. A primary type colony has been defined as "a colony of any laboratory animals defined genetically and of known nutritional and disease status. Its function is to provide breeding stock for subcultivation elsewhere. Also known as foundation stock" (Report, 1964, b).

The implication is that a primary type colony is small and that a great deal is known about it. Genetically it has, by rule, to be defined: not in its entirety, for that would be impossible, but at least in some relevant or useful respects. This is no less true of inbred animals than of outbred.

The reference to definition of nutritional status is not important, unless there is a real likelihood of the animals having some nutritional deficiency, which is not so common today when so many good compound diets are available (it was not always so).

"Disease status" is an unexpected term to use in such a definition, for it seems to imply that all laboratory animal colonies may be expected to show some signs of disease. But the definition was published in March 1964; today disease would be regarded as exceptional and unacceptable in most primary type colonies.

To conform with the definition some convincing degree of quality control has to be carried out. It became evident in the foregoing chapters on quality control that this cannot be done directly on large colonies without using in the process an unreasonably large number of animals for testing, which may still be an unconvincingly small sample of the whole colony. This applies particularly to health control. When it comes to inbred strains, many—indeed probably the great majority—do not exist except as small colonies, and some at least require quite laborious genetical methods to maintain them and preserve their more interesting characteristics.

For all these reasons a primary type colony must be small, perhaps fewer than 100 breeding animals. Because of the need for constant monitoring of quality, already mentioned, they must be maintained in centres where the facilities exist to carry out such monitoring, and in a way that will command general confidence. For if the monitoring is not carried out, or if it is done so inexpertly that it carries no conviction, the definition of a primary type colony cannot be complied with.

Thus primary type colonies will tend to be accumulated in recognized primary type colony centres, which are defined as places "where one or more primary type colonies are being maintained" (Report, 1964, b). Recently ICLA (the International Committee on Laboratory Animals) has begun to list recognized primary type colony centres, and those so listed are expected to carry general confidence. But there is nothing to prevent any other centres maintaining primary type colonies and carrying out effective quality control procedures on them, from being similarly regarded. It is really a question of confidence, and whether the centre can show convincingly that its primary type colony actually conforms to the definition given it.

III. Expansion

A defined strain, whether inbred or outbred, must originate in a primary type colony: but this, by definition almost, must be small. If the animals of a defined strain are needed in large numbers, larger than a primary type colony can furnish, a second stage of expansion has to be organized.

There are two parallel considerations in the organization of an expansion breeding unit: to ensure genetic conformity, and to control health. They are not incompatible or mutually exclusive, but they will be taken separately.

A. GENETIC CONFORMITY

In Chapter 4 it was seen that an inbred strain was not only liable to be rather lower than an outbred in fertility, but that to ensure genetic uniformity throughout the strain the development was to be avoided of parallel lines of breeding separated from each other by many generations from a common ancestor. There must be a main line of ancestry, with only short branches. But this precludes the production of large numbers; or if the same degree of documentation has to be carried out on large numbers as is necessary in the primary type colony, it makes it entirely uneconomical.

Some years ago a system was evolved at the Laboratory Animals Centre that offered a simple solution to this problem (Lane-Petter and Bloom, 1957). In this system the primary type colony centre provides small numbers of authentic breeding stock to an expansion unit, which is required to follow a simple but rigid programme of further breeding. This programme allows three, and only three, generations of subcultivation from the primary type colony: the green, the yellow and finally the red generation, at which point breeding stops. The system has become known as the traffic-light system, for obvious reasons.

Animals sent out from the primary colony are distinguished by a white label on the cage. The first, second and third generations of subcultivation are labelled respectively green, yellow and red. Progeny from white-label cages, when made up for breeding, are put into cages that carry a green label: those from green-label cages into yellow-label: and those from yellow-label cages into cages carrying a red label. Animals born in red-label cages are for experimental use only, and must never be bred from in any circumstances. When the subcultivation programme is in full operation, there will be a few green labels, more yellow, and a great preponderance of red. Before the greens disappear and the yellows are too depleted, more white-label animals must be obtained from the primary colony, to keep the cycle in continuous operation. Animals must always, under this

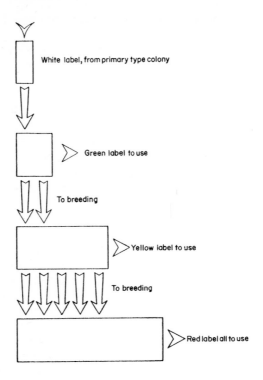

Fig. 10.1. In a production unit there will be a few pairs of white-label animals from the primary type colony: a small number of green-label and a larger number of yellow-label, the progeny of which will go mainly for breeding, but some of which may be used for experiment: and a majority of red-label whose progeny must all go for experiment and none for breeding.

system, be mated with those of their own colour: green with green, and yellow with yellow. The whole system is illustrated schematically in Fig. 10.1.

It is not necessary that experimental animals be drawn only from red-label cages. It will seldom or never be necessary, or even possible, to breed from all the animals born in green- or yellow-label cages, and those that are not used for breeding can go for experiment. This allows some scope for selection in the production unit, selection of a subjective and not too systematic kind, but none the less helpful in keeping up a good standard of production. The maximum ultimate production of animals will be thereby reduced, in proportion to the numbers not selected for breeding, but local needs will determine these proportions in the light of fluctuations in supply and demand.

Where inbred strains are being subcultivated by this method, there is no point in insisting on continued brother by sister mating. In three generations of random mating the loss of homozygosity will be negligible. It is therefore permissible, even desirable, in a production unit to treat inbred as if they were outbred, and to mate them without regard to relationship, provided that they are mated only to animals of the same generation. This will simplify recording in the production unit considerably.

It must be accepted that in the production unit things can go wrong. There is always the possibility of infection getting in, perhaps also of nutritional disasters, and even of accidental miscegenation. The likelihood of these things happening will depend largely on the care exercised within the production unit to avoid them. But care beyond a certain point is costly, and compromise has to be struck between the cost of extra care and the risk of trouble. The rules laid down by the primary centre should indicate which are obligatory—for example, the traffic-light code—and which are merely desirable—for example, certain refinements of hygiene, ventilation and the like. It is then up to those in charge of the production unit to balance the risks against the refinements. They have the consolation, if trouble does overtake them, that in the event of a total loss of their colony, they can easily start again with the same authentic strain. The interruptions will be temporary but not irrecoverable.

The traffic-light system of subcultivation of inbred strains can be operated by animal technicians who do not possess any special genetical skills or training. It is little if at all more laborious than any other system of breeding, and it gives the greatest probability of genetic uniformity in a large breeding unit. It does not give total certainty, of course; that is unattainable because of the occurrence of mutations, which will be more numerous in a large population than in a small. But the system does prevent a colony breaking up into a number of subcolonies which in time diverge more and more from each other. It, or some modification of it, is a practical way of breeding large numbers of inbred animals.

B. HEALTH CONTROL

There is an altogether different reason for having a primary type colony, other than to ensure genetic uniformity in a large production unit of inbred animals.

Consider a small colony of animals in an isolator. They are gnotobiotic; that is, they are demonstrably either germfree or associated with one or more designated microorganisms, and no others. Move these animals out of the isolator, but into an area where stringent barrier conditions obtain. They will acquire a number of microorganisms whose full inventory cannot be known in its entirety, and therefore they have ceased to be gnotobiotic. But if the barrier is good the inventory will not be large, and it should be limited to benign organisms: that is, organisms not usually associated with disease. But a barrier system is less absolutely exclusive than an isolator and must be regarded as an open system: not very wide open, but open just the same. No such barrier is able to ensure all the time the total exclusion of all foreign microorganisms, and in fact from time to time the barrier will be penetrated. When that happens the invading germs may or may not establish themselves inside the barrier, and they may or may not take up residence in the animals. Here they may or may not cause disease, and if they do cause it, it may or may not develop into an epidemic.

The point of all these hypothetical suppositions is that the traffic of germs is always one way: from outside to inside. No matter how slow it is—and a good barrier will mean that this traffic moves very slowly—new microorganisms will accumulate inside the barrier, become resident in the animals and, in time, cause epidemic disease. The accumulation is, for all practical purposes, irreversible. It is also cumulative, that is, new organisms are acquired step by step, so that the inventory is constantly being added to, even by benign organisms. An easily recognizable outbreak of serious disease may be an obvious turning point in the history of the colony, but in many cases the preceding months of silent acquisition of new organisms made the ultimate epidemic inevitable (see Chapter 5).

The better the barrier, the slower the invasion, but also the greater the trouble and expense of maintaining it. At some point, before the inevitable epidemic occurs, it would be nice to start all over again, from the isolator where, from the germ point of view, conditions do not change.

What are the time scales in this inexorable breakdown of barrier animal houses? Time is prolonged by strict barrier precautions, by small populations, by low densities, by excellent hygiene, by good ventilation and by all-round good husbandry. The breakdown is hastened by the converse of all these factors, and by bad luck. A large rat or mouse breeding unit of some thousands of breeders, in moderate to high density, with a strict barrier—showers, autoclave, etc.—may last 2, 3 or more years before a serious epidemic occurs. Smaller barriered units may last longer; others may break down within months. This time scale has a vital bearing on the control of health in expansion breeding units.

C. MULTI–STAGE BREEDING

When producing animals of an inbred strain, a small primary type colony has to be established for genetic reasons, in order to provide white-label breeding stock to the expansion unit, where they will be bred in large numbers by the traffic-light system or some variant of it.

For outbred animals it is important to maintain the maximum gene pool, and so the primary type colony may not be permissible in the same way as for an inbred strain. But this does not rule out the need for discontinuing the production colony at regular intervals.

For reasons of health a production unit cannot be continued indefinitely, whatever the nature of the peripheral barrier—other than an isolator—and in practical terms its useful life may not exceed 2 years.

A practical solution that will meet all these requirements is the following.

Animals for experiment are bred in a production unit, consisting of a number of not very large rooms—perhaps large enough to be looked after by one animal technician. Each room is furnished with a complement of breeding stock, which remain in that room for their total useful breeding lifetime, unless they are culled for some reason, or die. At the end of their breeding life time they are all destroyed, the room is emptied, disinfected and re-stocked with a further complement of breeders. The process is then repeated indefinitely.

In the case of mice or rats the useful breeding lifetime is about 9 months, which is well below the 2 years in which a barriered unit can begin to expect trouble. During 9 months, in effect, the accumulation of new microflora will not be likely to have reached a critical level.

The source of the replacement breeding stock is a primary colony of sorts. Not to confuse it with a primary type colony, let us call it a white colony. The function of this white colony is to produce breeding stock for the production rooms. In the case of rats and mice, the number of animals in the white colony needs to be about 5 per cent that of all the production units taken together. Thus the white colony is relatively small, and it can also have a more stringent peripheral barrier. These two facts alone are likely to prolong the period before eventual breakdown to more than the 2 or 3 years already mentioned, but ultimate breakdown may be expected in time. The white colony, therefore, needs turning over periodically, although less frequently than every 9 months.

The source of new breeding stock of the white colony has to be from an isolator: that is, they have to be gnotobiotic or ex-gnotobiotic animals. But they do not need to be of the same strain, because even if they were their numbers would be too small to maintain an adequate gene pool in the case of outbred animals: or in the case of inbred animals they would interrupt the line of inheritance. In either case, hysterectomy is a necessary part of the programme. For inbred animals, a single successful hysterectomy of a litter on the main line of inheritance, carried out from time to time, will ensure that a continuous line is cleaned up from time to time. For outbred animals, a number of dams from the production unit would have to be hysterectomized and their pups fostered on to dams in the white colony. In this way both cleaning up and conservation of the gene pool are effected (see Fig. 10.2, p. 222).

There are two points to notice about this system, as applied to the production of outbred animals. For genetic reasons a supply of pregnant dams

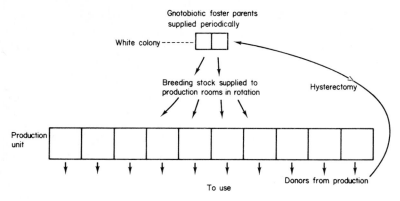

Fig. 10.2 Scheme of rotation of breeding stock, with genetic continuity and infective discontinuity. Each room in the white colony is periodically supplied with gnotobiotic or ex-gnotobiotic stock, which will be the foster parents of young delivered by hysterectomies of dams taken at large from the production unit. In this way the white colony produces breeding stock for transfer in rotation to production rooms. At any one time, one production room is building up, and another running down before being cleared out, fumigated and re-stocked from the white colony.

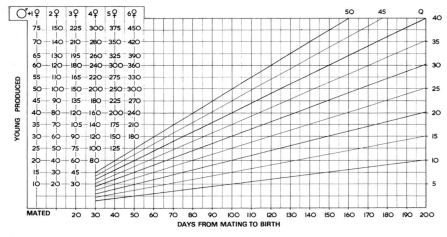

Fig. 10.3 Nomogram for calculating cumulative productivities in breeding groups, from pairs ($\male + \female$) to harems ($\male + 6\female$). To calculate Q, plot the point corresponding to the number of young produced against the number of days from mating to birth of the young (prenatal days). This will fall on or between the sloping lines, from which the value of Q (the number of young per hundred prenatal days) may be read.

is taken from the production rooms, selected from as wide a field as is practicable so that the gene pool of the colony is not diminished. These are hysterectomized and the pups fostered on to gnotobiotic or ex-gnotobiotic foster dams in the white colony. In this operation the genes come through, but the infections that have accumulated in the production colony are eliminated by the hysterectomy.

The second point of importance is that every room, both in the production unit and in the white colony, is periodically taken out of use. This enables each room to be regularly fumigated, as well as offering an opportunity for carrying out structural repairs, repainting and the like. If there are ten production rooms and the cleaning, repair and fumigation process takes 1 week, 10 weeks out of the operational lifetime of a roomful of breeders—say 40 weeks—will see only nine rooms producing. In addition to this 2·5 per cent loss in theoretical maximum output, a room that has been cleared out will take about 6 weeks to build up to full production, and a room about to be cleared out will need up to 6 weeks to run down. (These figures apply to rats and mice and represent the sum of a gestation and a lactation period.) Attempts to shorten this period of phasing in are bound to lead to uneven production in subsequent weeks.

The phasing in and phasing out will lead to a further 10 per cent loss in maximum theoretical output of a production unit, making a total of 12·5 per cent. It will be a continuing loss, for all time, but it will be less costly than periodical breakdowns that inevitably occur sooner or later in a unit that is not turned over in this way.

In the white colony re-founding of each room is just as necessary, but at less frequent intervals. The turnover period is 100–150 weeks, instead of 40, but the build-up from gnotobiotic importations, together with the inevitable wastage associated with hysterectomy, will lengthen the phasing-in and phasing-out periods. However, in the white colony productivity per unit floor area is of much less importance than in the production unit, because the white colony is so much smaller.

For inbred strains the procedure of assuring genetic continuity with infective discontinuity is essentially the same, except that the white colony is very much smaller than an outbred colony. The production unit of inbreds is also likely to be much smaller. It may be found more convenient to hysterectomize the inbred animals into an isolator, which takes up very little space, rather than into a room, which cannot accept any other animals apart from those that are born there. During the first 3 months or so in the history of a newly founded inbred colony the numbers of animals are so low that they can be housed in one or two isolators. The animals do not have to remain germfree or strictly gnotobiotic during their period in the isolators; the isolator is regarded purely as a small room, and only serves to avoid under-using a room in the white colony during the early weeks of build-up of an inbred strain deriving from a single pair of animals.

D. DISPOSITION

The general pattern of a laboratory animal breeding unit is, therefore, set by the foregoing considerations. The very small unit poses no special problems, because almost any approach to the task of breeding will produce animals, and the efficiency of the unit is not a matter of great moment.

It might also be thought that the cycle of re-derivation, through hysterectomy, would only apply in special cases, where large numbers, high densities and exacting health standards were demanded. But this outlook is becoming less and less tenable, because of the rising costs of animal accommodation, the economic need for efficient methods, and the more exacting health standards that most users require. If it can be shown that for example, rats in large numbers free of respiratory disease can be produced at a price little if any above diseased rats, then the healthier animals will be used.

But this chapter is concerned with breeding units of moderate or large size, from which will come substantial numbers of animals in standardized groups—so many of one sex at a given age or weight range. The operation of a unit of this kind demands a high degree of predictability of output, as well as the exercise of continuing ingenuity in matching supply and demand.

The alternative to this is, on the one hand, disappointed or dissatisfied users and, on the other, the regular destruction of animals that have been bred but do not fall into the appropriate weight or age groups at the time they are wanted. This destruction can occur simultaneously with specific demands not being met for lack of suitable animals.

From the point of view of the health of the animals, and of simple control of a breeding programme, small rooms, each absorbing the work of a single technician, are preferred. From the point of view of meeting supply and demand, large rooms are much easier to operate. But with a suitable system of recording the availability of animals, the operational drawbacks of smaller rooms can be overcome. In view of the need to turn over breeding stock in production units at regular intervals, it must be conceded that the day of the large breeding hall, at least for rats and mice, is over. Suites of one-technician rooms must replace assembly halls.

In practice, it may be found inconvenient to allocate one room to one technician, despite the soundness of the general principle. There are occasions when the technician is away, on the sick list or on holiday, or the colony is reduced in size and does not provide a whole day's work. There is also the psychological effect of working continuously in the exclusive company of the animals; young people in particular are subject to this sort of loneliness. It will, therefore, be better to allocate, say, two rooms to two technicians, so that, even while each takes primary responsibility for one room, they can work together, and one can stop in whenever the other is absent.

In the process of emtying a room, cleaning it up and re-stocking it, that room will be out of production for a period. The length of this period will depend on precisely how the replacement is organized, but it can never be less than a week, and may exceed a month. Therefore, in any plan, account must be taken of time, and it follows that the more the rooms the less the penalty to be paid for this regular interruption in their utilization. The production of one room will, in fact, be run down, allowing the last series of pregnancies to go to term and the young to be raised, at the same time that another room, established with new breeding stock, is building up to full production. This overlap period will be about 6 weeks in the case of rats and mice.

The penalty for interruption of production rises sharply as the number of breeding rooms falls short of ten.

The same principle applies, but on a smaller scale, to the white colony. For rough calculations, in respect of rats and mice, the shelf space allotted to the white colony will need to be about 5 per cent of that allowed to the production unit; but the floor space will be greater than this, because the white colony is in smaller rooms in which the floor space can be used less efficiently in relation to available shelf space.

If one animal technician can look after a room of about 40 m², containing about 200 m of shelf run, this would form a useful size of production room. For the white colony, an area of about 4 m² with some 10 m of shelf run would support one such production room. This would be an unreasonably small room, so that one of, say, 8 m² with 20–30 m of shelf run would be more practical. Since the white colony is small, and the rate of replacing breeding stock in it is likely to be much less frequent than in the production unit, a satisfactory compromise would be achieved with much less than ten rooms.

In summary, then, a substantial breeding unit would consist of a number of production rooms, each of a size to be cared for by one animal technician, and if possible not less than ten in number. To support them there would be a white colony, of total area about 10 per cent of the production unit, divided up into much smaller rooms, each 10–20 per cent of one production room, and in number not less than about four. That would be a typical disposition of production facilities, together with the communication and areas discussed in Chapter 6.

IV. Breeding Systems

There is a true story of a certain head of a laboratory, who decided that, since about 50 per cent of the embryonated eggs he was buying were found to be infertile, he would produce his own. He had the land and the staff to do this. So he bought some chickens, grew them on, and waited for the eggs to be laid. Unfortunately, the fertility rate proved to be even less than 50 per cent; it was nil. The laboratory director, having had no one to advise him, had omitted to install a cock in his hen run.

The moral of this story is that laboratory animal breeding is not so simple as it seems. It is better to consult an expert about how to do it, or better still to leave it to those who really know about it. The results may not be perfect, but they will probably be better than the director's.

There are systems of animal breeding that must be devised and followed if good results are to be obtained. All laboratory rodents have a post-partum oestrus, and are spontaneous ovulators; ovulation can also be induced in rats (see Ritchie and Humphrey, 1970). It is possible to breed from them all at the onset of sexual maturity, but if this is done the first litters are likely to be disappointing. On the other hand, if first breeding is left until adult size is reached, the animals may well be rather fat, and their fertility will be lowered.

After several litters the productivity of the animals will be at its height. The productivity of the females can be measured in terms of the number of young produced in a given time, per week, or per 100 days, for example. The productivity of the male can only be assessed in terms of the number of females he gets pregnant in a given time. It is often found that male fertility falls off at an earlier age than females, possibly due to the fact that the males in breeding colonies tend to get too fat.

When the individual breeder reaches its maximum productivity it should be culled, for thereafter the productivity will fall and the animal will not earn its keep. For any given strain and set of environmental conditions this point of optimum culling of breeders should be ascertained by experiment. The fact that a few breeders could last longer should not tempt one to keep them on, for to do so will disrupt a breeding programme, and in the long run will not pay.

Table 10.1 gives some vital statistics useful for calculating productivities

Table 10.1

Some representative vital statistics relating to the breeding of the commoner species of laboratory animals (compiled from various sources)

	Mouse	Rat	Hamster	Guinea-pig	Rabbit	Cat	Dog
Gestation period (days)	17–21	19–22	16	62–68	28–31	64–66	62–67
Oestrous cycle (days)	4–5	4–5	4	14–16	[a]	14	5–12 months
Age at first mating (weeks)	6	8	6–8	12–15	25–40	30–40	60
Age at weaning (days)	18	21	21	7–21	42–56	42	42–56
Average litter size	12	12–14	10	3–5	5–12	3–6	4–8
Age of retiring breeders (months)	9	9	12	18	24	48–60	<120
Productivity per ♀ per week	1·5–2·5	1·5–2	1·2				
per 100 days (Q)	21–35	21–28	17				
per year				16–20	30	8	8

[a] The rabbit is an induced ovulator

in some of the common species of laboratory animals. Since these figures vary widely, with the strain and with the circumstances in which the animals are kept, they should be used only as a guide to estimating production. For example, inbred strains of mice will not achieve the productivity shown, but some strains of mice or rats may do better than the figures indicate. Fuller vital statistics will be found in tables in Short and Woodnott (1969).

A. INBRED STRAINS

For all practical purposes, inbred strains must be bred in monogamous pairs, the male and the female being mated at or some time after reaching sexual maturity, and remaining together for the rest of their lives. (In some species this may not be possible, because of fighting, but it does not occur with the common laboratory rodents.) The young are born and removed when weaned. In a proportion of cases the male will impregnate the female within hours of having her litter, at the post-partum oestrus, and so she will sometimes be in simultaneous pregnancy and lactation.

The percentage of successful post-partum matings is never 100 per cent, and often as low as 50 per cent or lower. Among the reasons for this are failure to mate; male infertility; failure of the female to implant the ovum; pseudopregnancy; and resorption of the implanted ovum. There is frequently a delayed implantation if the female is simultaneously pregnant and lactating, and this in the mouse may be as long as 10 days.

At the end of lactation the female will come into oestrus within about 3 days in the mouse and rat, and accept the male. If she is not then mated, she will come into oestrus regularly until she is mated. A successful mating will be followed by a new pregnancy, and an infertile mating by a pseudopregnancy, which will last for about 14 days in the mouse or the rat.

Monogamous pairs make a running calculation of productivity easy. The productivity of a female may be defined as the number of young born and weaned in unit time. Unit time can be a week, 100 days, a year or any other convenient period. The time is counted from the day the pair is first mated to the birth of a litter: these are called the prenatal days. The calculation cannot be made until the litter is weaned. The calculation may be intermittent: that is, it can be made anew for each litter produced. Or it can be cumulative, summing all the prenatal days and all the young produced. The cumulative calculation is more useful. The calculation is represented by the formula:

$$\frac{L \times 7}{D} = P_W \quad \text{or} \quad \frac{L \times 100}{D} = Q$$

where L = young produced during the period, D the number of days in the period, P_W the productivity per week and Q the productivity per 100 days (Lane-Petter, Brown, Cook, Porter and Tuffery, 1959).

The figure P_W is more useful for calculating the output of a colony, but it requires two arithmetical operations on a slide rule: divide L by D, and multiply the quotient by 7. The figure Q requires only one operation, namely dividing L by D, and this can save trouble if large numbers of calculations have to be made. However, an easier way is to use a calculating nomogram, such as is shown in Fig. 10.3. Such a nomogram is easy to construct for any points of reference.

In pair mating, a failure to achieve a reasonable level of productivity can by this means becomes immediately evident; as can a falling off of productivity. Low values should ordinarily lead to culling of the pair (except when a very sub-fertile inbred strain is concerned, when even low productivity has to be conserved). They may also give warning that there is a cause for the low productivity—either infective, nutritional or something else—and this should lead to investigation.

The running calculation of productivity in a pair-mated colony is also a useful criterion on which to base a selection programme. In general, selection is not permitted in inbred strains; that is, selection that would tend to transform the character of the strain in the course of time. But viability of the colony depends on maintaining a certain level of productivity, and this is often only possible, and almost always highly desirable, if selection for productivity is practised. The sort of selection will tend to hinge on factors that are polygenically controlled—that is, depend on characteristics that are influenced by a number of genes—and there is little or no scope for this in an inbred strain. But it will also pick out characteristics that are non-genetically passed from parent to offspring, and one such characteristic is infection.

In summary, monogamous pairs are most suitable for inbred strains. The calculation of productivity is simple and can be continuous, thus providing a ready means of selection.

B. OUTBRED COLONIES

Breeding systems for outbred colonies may be divided into two groups: those that attempt to utilize the post-partum oestrus for regular mating, and those that do not. The former may be called continuous systems, because there is no theoretical reason why the female should ever be empty, except for an interval of not more than a few hours between parturition and the start of a new pregnancy. The second group of systems entails separating the female from the male at least for the period of a post-partum oestrus, and usually for the duration of lactation, so that the female is not pregnant during lactation or until she is mated at a post-lactational oestrus. Such systems may be called discontinuous.

1. Continuous Mating Systems

The monogamous pair system described in the previous section as most conventional for inbred strains may equally be used for outbred colonies.

All the advantages of easy running calculation of productivity that were described are also present with outbred colonies, and recording may be just as for inbred strains. Selection will be effective in an outbred colony because there is much more heterozygosity present to form a field of selection, and very high levels of productivity, in terms of young produced per breeding female, may be achieved in this way. However, it must be remembered that for every breeding female of a monogamous pair there is a breeding male, and if the productivity is given as the number of young produced per breeding animal, male and female, the figure will be only half the value, and not very impressive.

Continuous breeding may also be achieved by placing one male permanently with two or more females, in trios ($1\male + 2\female$), quartets ($1\male + 3\female$) or harems ($1\male + n\female$). The value of n is only limited by the ability of the male to impregnate a number of females; and, of course, by the size of the cage, which may itself have some influence on the mating capacity of the male. Some breeders have made up harems containing two males and a greater number of females, but it is always found that one male is dominant and does most or all of the mating, while the other, if it does any mating at all, certainly does not earn its keep. With rats and mice, a harem of $1\male + 9\female$ is near the practical limit of the male's mating capacity. With guinea-pigs a harem of $1\male + 12\female$ is perfectly feasible, and the capacity of the male is able to cope with up to 20 females.

A common size of harem for mice is $1\male + 6\female$, the male being left in the harem all the time. It is found, however, that with an increase in the number of females, the mortality of the pups rises, and the weaning rate falls, especially when there are three or more females; at six females it may be serious. Thus a trio of $1\male + 2\female$ will not produce quite twice the number of young as a pair; a quartet less than three times; and a harem of $1\male + 6\female$ considerably less than six times. The productivity per female will fall, but so will the total number of breeders, male and female, and on this basis the disadvantage of larger harems is not so great.

Causes of loss of young in harems are the overlaying of the younger pups by older ones in the same nest; the stripping of milk from dams that have just had young, and should be feeding them, by older pups from other dams, leading to the starvation of the younger pups, and stampeding in the cage of the adults, leading to injury and death of the young.

On the other hand, continuous breeding is easy. Once a mating group is made up, it has only to be fed, watered and cleaned out, and the young harvested when they appear to be ready for weaning. Birth dates are not recorded, nor individual litters, nor the performance of individual males or females. Weaning age is approximate only. Only the overall production of the harem, and the total prenatal days, can be recorded and used to calculate the productivity, and this will be shown on a harem basis, thus severely limiting the opportunities for selective breeding. It is a simple system, like backyard poultry keeping, but it cannot today be regarded as compatible with good standards of breeding and husbandry.

Rats may be mated in permanent monogamous pairs, but after some weeks the male rat grows so big, and eats so much, that the system has little of practical value to recommend it. Rats will breed in permanent harems, but not with the same facility as mice, and continuous systems with rats are not much used because of high weaning losses.

Guinea-pigs are particularly suited to permanent harems, for they are very gregarious animals, and they make no nests. A male and from six to twenty females will form a breeding group, which may be left together for $1\frac{1}{2}$ to 2 years, the young being harvested at about 2 weeks of age. It is comparatively easy to know actual birth dates of young, because there are not so many to record: it is also not difficult to record the performance of each dam, except when two in the same pen or cage litter down together. Thus selection at an individual level can be practised. The long gestation period of the guinea-pig (about 9 weeks) and the short duration of lactation (about 2 weeks) mean that simultaneous pregnancy and lactation do not impose an undue strain on the dam.

To summarize, continuous breeding systems include monogamous pair mating, most useful for inbred strains, but applicable, if sometimes inefficient, to outbred colonies of mice, rats and guinea-pigs. Trios, quartets and larger harems may be used for mice, with moderate efficiency, although large harems incur too many juvenile losses. They are not suitable for rats, because of heavy juvenile losses; but large harems work well with guinea-pigs. In calculating productivity, it is more realistic to include the male breeders with the females, and to reckon the number of young produced in unit time per breeder, not just per female.

2. Discontinuous Mating Systems

In discontinuous mating systems, the female is separated from the male some time between mating and parturition, so that post-partum mating cannot take place. A new pregnancy cannot be started until the post-lactational oestrus at the earliest.

In rabbit breeding it is normal to put the female to the male, who will then mate her immediately, the female rabbit having no regular oestrous cycle, ovulation being induced by the act of mating. The female is then replaced in her cage to await the progress of pregnancy. That is one extreme.

On the other hand, a mating group of one male rat or mouse and a number of females may be left together until the females show advanced pregnancies, and when they are judged to be a day or two away from term they are removed to maternity quarters. In this way the greater part of pregnancy is spent in the harem, which will lead to some economy of space. At the end of lactation the females are returned to the male, to be mated and start another pregnancy.

The maternity quarters, in which the females are placed just before parturition, may be single cages, which are suitable for rats, or communal cages containing two or more parturient females, which are suitable for mice. Rats, in fact, do not mix well together when they have young, although they can be persuaded to tolerate each other. On the other hand, mice do not raise their young so well when the females are on their own as they do when there is a group together. A practical group is three females, and their progeny, all of the same age, so that there is no deprivation of the younger pups by older. All the young will be collected in the same nest, and suckled indiscriminately by all three dams.

In discontinuous systems individual recording presents no problems, and thus a selection programme is always easy to apply. The loss of productivity, when compared with a continuous system, is less than might be expected, for two reasons. First, despite the presence of a potent fertile male, not every post-partum mating will lead to pregnancy: the rate is almost never above 80 per cent and is usually much less. Second, there is very often a delayed implantation, so that even with successful post-partum mating the interval until the next litter is born will be some days longer than a normal gestation period. Thus, the sacrifice of the post-partum mating is not such a great loss as it might appear.

3. Cross-Fostering

The systems of breeding so far described have all entailed leaving the young that are born with their natural dams; either exclusively, when the dam is in a single maternity cage, or collectively, when there are two or more dams together. Now, in all colonies, including those that are strictly inbred, litter size as born varies within wide limits. Large litters may be raised with negligible losses, but there will be an inverse relationship between the size of the litter and the individual weights of young at weaning. Because litter sizes at birth vary so much, weaning weights will also vary, and this will lead to considerable disuniformity of weight for age, which will persist long after weaning and perhaps throughout life. Standardizing the size of litters would cut out this cause of lack of uniformity.

It has long been common practice to foster excess pups from very large litters on to very small litters of the same age if they are available, or to kill the excess pups if they are not. A systematic way of doing this with rats was described by Lane-Petter, Lane-Petter and Bowtell (1968), who showed certain practical advantages of the system.

Briefly, cross-fostering is a method of raising rats or mice in litters of standard size. It is only feasible in rather large breeding units, in which a substantial number of litters are born each day; for example, not less than 1000 ♀ breeding mice or 500 ♀ breeding rats. From litters that are all 2 days old the young are sorted out into groups of the same size, and these groups are given back to the dams from which they came. Some at least of the pups that go to each dam are not those she bore, but she will nevertheless raise them as her own. In the case of rats, where the sex is easily seen at 2 days, each group may be all of one sex, and if the demand from the breeding colony is for a preponderance of one sex the surplus pups of the unwanted sex may be destroyed at 2 days instead of being grown on to weaning. If there are not enough pups to give each dam a full complement, the surplus dams may be returned to the male, when they will come into oestrus within a few days.

The litter size as born may well be less than the litter size that a dam can successfully rear. This, together with the removal at cross-fostering of surplus pups of the less popular sex, will lead to a surplus of dams, and thus reduce the average number of prenatal days within the colony.

The above system works well with rats, but a modification of it is necessary for mice. To begin with, sexing 2-day-old mice is much more difficult than rats, and unless there is a particular reason for it, it is seldom worth the trouble. Also, mice, even on a discontinuous system of breeding, do better in groups of dams with their litters. A group of three dams does well, and they may have between them as many as thirty-six young to raise. Apart from these considerations, the technique of cross-fostering mice is much the same as rats.

The advantages that accrue from this technique are a high degree of uniformity of weight for age at weaning, leading to a high predictability of the output of given weights on given dates; counting of the young in nest is much easier because each nest contains a standard number of young; and great economy in the use of breeders, because all females are used to the best of their lactational ability and all males to the best of their mating ability.

Technically, cross-fostering is easy to learn and teach. It occupies a little extra time at cross-fostering, but it saves much more time at weaning, and subsequently, because all pups are of standard size and the weight scatter is so much reduced. Where such considerations are of importance, cross-fostering is a technique that cannot be ignored.

4. Breeder Replacement

It is uneconomic to keep breeding males or females after they have passed the peak of their productivity. They should therefore be replaced.

The rate of replacement is a function of the number of breeders ($\male$ or $\female$) in the colony, N; and the average breeding life time of X weeks. The average weekly replacement rate, R, is then given by the formula

$$R = \frac{N}{X}$$

The value of X will not necessarily be the same for males as for females.

Account must also be taken of wastage of breeders through culling sub-fertile or infertile animals, or those which are in any way inferior.

Suppose that in a colony of 520 breeding females the useful breeding life time is 26 weeks, the average weekly replacement rate will be 520/26 = 20 females. To this must be added replacements for culled animals; the sum of those two will have to be selected and set aside from stock.

This is the situation of a colony that continues to breed indefinitely, although the breeders are being constantly replaced. However, in the more practical situation, when a breeding colony is set up in a room, and at the end of its breeding life, X weeks, is destroyed and, after cleaning up, re-placed completely, the replacement breeders will have to be found in a shorter period of time.

In replacing a whole colony of breeders it is, in fact, more practical to spread the replacement over some 3 weeks, because this will ensure a steady flow of young being born, instead of a brief population explosion followed by a period of zero births.

C. ECONOMIC CONSIDERATIONS

In choosing a system of breeding some regard will be paid to economic considerations. Given an animal room of a certain size, how many rats, mice, guinea-pigs, can be produced from it? This is a question of produc-tivity, and productivity may be measured along different scales.

The productivity of a female is the number of young she can produce in a given time: according to the formula given on p. 227:

$$P_W = \frac{L \times 7}{D} \quad \text{or} \quad Q = \frac{L \times 100}{D}$$

where P_W = number of young per 7, or Q per 100, prenatal days, L = number of young and D the number of prenatal days. For successful breeding, selection for this sort of high productivity is essential, and any system that enables it to be done on this basis of individual animals will lead to a more productive colony than where it can only be done on a group basis (as in permanent harems of mice, for example). Whatever the system of breeding, selection for individual productivity is of overriding value.

From the point of view of economy, however, productivity per breeding female must be set against other aspects of productivity. Pair mating re-quires as many breeding males as females, and these all take up room and eat food. Polygamy has an advantage.

Pair mating also requires more cages, which occupy more space. In fact, pair mating may produce the largest number of young per female in a given time, but it will occupy the greatest footage of shelving in the animal room, and so is uneconomic in space. On the other hand, discontinuous mating systems are very economical in space, because of the small number of breeding males needed. In the use of a discontinuous system with rats and mice, at any time the number of females in maternity cages will roughly equal the number of females in the mating cage with the male. Thus, if one male can manage eight females, there will be eight more females raising young: a proportion in the colony of $1\male : 16\female$. With a continuously mated harem, the colony proportion would be only $1\male : 8\female$, and the weaning losses in such colonies are always much higher than when the females are placed in maternity cages of the right size and, in the case of mice, with the right number of other females.

Then there is the influence of mating systems on the labour required. Pair mating absorbs most labour. Permanent harems are economical in labour, when measured against the number of breeders, though not necessarily against the number of young produced. Cross-fostering certainly economizes in labour, and for rats and mice is probably the most economical system of all.

Economic considerations, then, have to be related to productivity, and there is, in fact, a close relationship between economy of breeding and productivity. Productivity is measured by the number of young per female in unit time; per foot of shelf; and per pair of hands employed. No one criterion should be considered to the exclusion of the other two.

V. Critical Sizes of Breeding Units

It is a general principle of the hygiene of populations that the chances of infection becoming established in a colony of animals are increased, and the chances of eliminating such infections reduced, when the colony is large, the density is high, the traffic of animals within the colony is heavy, the traffic of animals and of human and other vectors into the colony is high, the standard of cleanliness is low, and when insufficient precautions are taken to ensure that food, bedding, water, air and other materials going in are not dangerously contaminated.

All of these factors except the last two, namely cleanliness and precautions about materials, are of more moment in larger populations than in small, so that mere increased size of a colony carries with it greater hazards of disease. From this point of view, then, a multiplicity of small breeding units would be safer than one large one.

On the other hand, the demand for large numbers of animals within narrow limits of weight and age on any particular day, and of as great a degree of uniformity as possible, cannot be met by such a multiplicity: large breeding units are essential. Such demands are particularly common for rats and mice, but they may also occur for guinea-pigs and rabbits, and also for chickens. There is thus a conflict between the relative security of small colonies and the operational convenience of large. There has to be a compromise, and the point at which the compromise is made, in terms of size of colony, will determine the health status of the animals coming from it.

A. POPULATION SIZE

The greater the total number of animals in a population, the more frequent will be the contacts between them: either direct, or by propinquity leading to airborne contacts. This effect can be mitigated if the overall population is broken down into smaller fractions, each fraction being treated as far as possible as a separate population.

Total separation of such fractions is not operationally possible for a variety of reasons. Since demands for large groups of closely similar animals—say, 5000 female mice, 20–25 g and within 5 days of the same age—cannot come from small colonies, it will require several such colonies to meet such demands. It is not difficult to order an environment and a husbandry routine that is the same, within narrow limits, for a number of small colonies, but the effect of different members of staff in each, and the subtle influences that they will exercise in each colony, cannot be entirely eliminated: they may, however, not be too important.

But if such colonies remained permanently isolated, one from the other, they will undergo different genetical changes which can in time be serious enough to introduce an unacceptable degree of genetic disuniformity. The 5000 mice will not be all alike. There must, therefore, be some integration of the overall population.

B. DENSITY

In addition to absolute population size, the density of animals in a large population is likely to be high.

To begin with, modern animal house accommodation is expensive and there are always strong economic pressures to make the most effective use of it. All animal houses must give up a proportion of their available space to non-productive functions, such as corridors, store rooms, offices, washing areas, etc; but the proportion of directly productive area, namely animal rooms, to the total animal house rises with the overall size. Therefore the density, expressed as the total number of animals in the animal house (not just the animal room) rises with overall size.

Density may be reckoned in other ways. There is the density of animals in the cage, measured as the number of animals, or the weight of animals, per unit area of the floor of the cage, or per unit volume in the cage. The number of animals rather than their combined weights is a more useful measure, and floor area is more useful than volume, but neither weight of animals nor volume of cage, which is a function of height as well as of floor area, should be ignored. A cage that is satisfactory for 30 weanling rats at 21 days of age will only hold 20 rats at 28 days, and 15 rats at 35 days. A kilogramme of mice will need more space than a kilogramme of rats. Optimal densities can only be determined by experiment; densities can be too low for optimal growth, as well as too high for health.

There is also the density of cages within the room. At first sight it might appear that this is irrelevant, for the animals are only concerned with the physical limits of their environment, which are set by the cage. But more cages mean more animals breathing, excreting and creating odours. In all animal rooms with conventional methods of ventilation this will mean interchange of exhalations and dusts among all the animals in the room: the higher the density of cages, the greater the interchange.

The use of filter racks, described on pages 131–132, to a great extent cuts out this interchange, in much the same way as would a lowering of the cage density or increasing the number of air changes in the room (provided the air is well distributed within the room). This approach, therefore, offers one way of counteracting the bad effects of high densities.

Reducing the cage density will result in less economical use of the available space in the animal house. It will also increase the number of steps taken, in the course of a day, by the animal technicians in their work; and the distances that have to be covered in bringing food, water, bedding and clean cages to the animals, and removing cages and soil. Once again, a compromise has to be struck.

C. INTERNAL TRAFFIC

An animal colony is not a static population, like a museum. It is constantly changing. Animals must be put together for mating, perhaps separated or reconstituted into new groups for littering, for growing on after weaning, for sorting out into new groups again that are required by the user, and so on. There may be deaths and culls, which need to be replaced. Such internal traffic of animals is essential to any production unit, and in large areas it can be very extensive. The more extensive it is, the greater the chance of a dangerous infection being rapidly and widely disseminated throughout the whole population. Conversely, small colonies limit such internal traffic, and thus circumscribe its dangers.

One of the more powerful arguments in favour of the system of multi-stage breeding described earlier in this chapter is that it makes possible this circumscription of infective hazards, while permitting a group of such circumscribed sub-units to operate as a single colony.

D. EXTERNAL TRAFFIC AND VECTORS

To bring an animal from outside into a breeding unit is extremely dangerous, because the microflora of the incoming animal cannot, unless it is gnotobiotic, be known in its entirety, and it may carry organisms that will provoke an epidemic. Only gnotobiotic animals may be safely introduced, or those so near to a gnotobiotic origin that their microflora is, for all practical purposes, known. In most cases the deliberate introduction of animals from outside must be totally excluded, or allowed only by the routine of hysterectomy and fostering. But accidental introductions are another story.

In any animal house that is not a strict germfree isolator, staff and materials that have not been sterilized are constantly coming in. A peripheral barrier is, apart from isolators, not an absolute bar to germs, but only a discouragement. It is likely to be penetrated from time to time by dust particles, even by insects, and it is not unknown for wild rodents to get in (though few who have charge of a new animal house would care to admit it if it ever happened). The food may be sterilized, but more frequently it is only pasteurized; and the same goes for the bedding and perhaps other materials. All these are possible vectors of infection: and in large units, more materials come in, the peripheral barrier is longer, and the chances of penetration rise proportionately.

Lastly there are the staff: both the regular animal technicians, and from time to time visitors. In many ways these may represent the greatest hazard of all, and the magnitude of the hazard is directly related to the number of entries that are made into the breeding unit. Quite apart from the special hazards of human vectors, by reason of contacts with animals elsewhere they may be a source of infection, the human vector hazard is directly related to size of unit—another argument for keeping breeding units, or their discrete fractions, small.

E. INTEGRATION

Indeed, there are so many arguments in favour of the small breeding unit, on grounds of health and manageability, that the case for large units has to be critically examined. The large unit is easier to plan and operate, because whatever the number of animals demanded from it, the process of meeting and making up a consignment is the same. They all come from the same assembly hall. Once the large unit is broken down into fractions, large consignments cannot come all from one fraction, but will have to be made up from the output of two or more.

This introduces a complication, for it has to be decided how many animals are to come from each fraction; and this in turn depends on the availability of animals of the right sex, weight and age in each fraction. But this is an administrative rather than an operational problem, and can be handled accordingly. There is too often a tendency to bend operational or technical needs to suit administrative convenience. It must be tempting to sit in an office and devise methods for others to practise, knowing that the blame for failure will be laid at the door of the operational staff. But this temptation must be resisted from the top down. Integrating the output of a number of fractional breeding units calls for some ingenuity, perhaps, but it is not so very difficult: it may be administratively less tidy than handling a single large unit, but there is much satisfaction in working out a good system of integration and seeing it work smoothly.

By integration in this sense is meant maintaining a correct tally of animals available in each functional breeding unit, broken down by sex and weight and age, and keeping this tally up to date at all times. In this way the inconvenience of smaller breeding units is largely if not entirely offset. The solution is so simple that objections to it only cause surprise.

The important question is, how big should these fractional subunits be? In an earlier section of this chapter it was suggested that animal rooms could conveniently be of such a size that one animal technician was needed to look after each. A smaller room may sometimes be needed: for white colonies, for special conditions of breeding, and for much experimental work: and these considerations will override the one room/one technician principle. Conversely, for work which habitually needs two technicians working together, a two-technician room may be justifiable. But these are exceptions: the one-technician rooms should be taken as the norm.

From time to time such a room will be taken out of use, because the breeding stock requires replacement, the experiment has come to an end, or there is an epidemic. If there is not to be a major interruption of work, there should be several other such rooms that can carry on while the one out of action is being cleaned up, re-stocked and brought into operation again. Perhaps up to ten such rooms can be run together: fewer, if that is the limit of need, but seldom more. If more are needed, consideration must be given to starting a second group of rooms, working independently of the first, rather than making the first too big for safety.

This suggests that there may be a critical size of animal house, above which there is a disproportionate rise in the risks inherent in size, and below which the gains in safety are balanced or exceeded by growing inconvenience or operational inefficiency. It is not possible to suggest a critical size in terms of absolute area, or number of animals, because so many factors affect the relation of these to each other, and of both to other considerations. But a guide is suggested by the relationship of an ideal group of rooms to its staff. Ten rooms, one animal technician per room, plus supervisors and ancillary staff of three or four, make a convenient unit of thirteen or fourteen persons, at the most. Five rooms might add up to seven staff, and would work well as a smaller unit.

This is as near as one can get to laying down critical sizes of animal houses, and it may be none too near the truth. But to consider the idea of a critical size is prudent, for it may prevent the building of large and unmanageable animal factories that are doomed from the beginning to failure.

VI. Distribution

Wherever laboratory animals are produced, they have to be taken to the place where they are to be used. This entails packing them in suitable containers for the transfer—which may be their normal cage for short journeys —carrying them over greater or shorter distances, unpacking them at the destination, and settling them down in new quarters. The majority of laboratory animals that undertake this journey are sent at or shortly after weaning, up to the age of puberty : it is at weaning and puberty that they are already under considerable physiological stress.

A. DISTANCES

It has often been observed that animals carried from the breeding room to the experimental rooms on the same floor may be as disturbed by the change of environment as those that have travelled considerable distances. This seems contrary to the suggestion that, to avoid translocation stress, animals should as far as possible be bred where they are to be used. It would perhaps be better to say that animals properly packed and transported suffer no more disturbances than if they were moved from one room to another.

Distance, then, is a small problem, and should be measured in hours rather than in miles. Journeys up to 6 hours from packing to unpacking present few problems, for there is no need to provide food or water for the journey, and the bedding will not have time to become foul. For most small laboratory animals, journeys up to 12 hours are much the same, except that some food should be provided, and the supply of absorbent bedding should be generous: no water is needed. Up to 24 hours, it is necessary to give water to rabbits and guinea-pigs, but not to rats or mice; but a succulent such as apple, potato or manufactured canned supplement may be offered.

For long journeys ample food is necessary, and succulent will certainly help to prevent weight loss on the journey, although rats can go 48 hours and mice 72 hours without water with little inconvenience. On arrival they will be somewhat dehydrated, but they will quickly restore their water balance when offered water, and be none the worse. They will quickly make up any weight lost through not eating.

B. PRIVATE TRANSPORT

Vehicles specially adapted for the carriage of laboratory animals, and making this journey from breeding unit to laboratory all the way by road, provide the best method of delivery.

It is important to remember that a truck load of animals represents a far higher density of cages per unit volume than any conditions existing in an animal house, and that the production of heat by such a concentration of animals is considerable. Although in cold weather some heating will be needed for the truck, the greater problem is to remove the animals' own heat, by means of an adequate circulation of fresh air. Losses from cold are much rarer than losses from heat.

A vehicle specially adapted for carrying laboratory animals should have, first, an adequate ventilation system. This should provide a steady stream of fresh air, at the rate of at least twenty changes an hour, entering and leaving the load space at a number of points, so that even ventilation is achieved in every part of it. The siting of the air inlets and outlets should be so arranged that draughts are avoided: both may need to be carefully baffled.

The air should be taken into the vehicle from roof level, to minimize the danger of sucking in exhaust fumes from other vehicles. It should be heated or cooled as necessary, and if possible fully air-conditioned. It is useful to place a thermometer bulb in an appropriate place in the load space, with a large dial on the dashboard of the cab, so that the driver can keep a check on the temperature to which the animals are being subjected.

The ventilation system must continue to operate when the vehicle is at rest and the engine switched off. This may mean fitting the vehicle with an extra battery.

The load space must be well insulated, and lined with sheet metal or some other material that is easily cleaned. Frequent washing down and spraying of the interior are necessary. An inner lining of sheet aluminium, with glass fibre insulation between it and the outer skin of the load space, has been found to work very well.

C. PUBLIC TRANSPORT

1. Road Transport

Public road transport is not suitable for the carriage of laboratory animals.

2. Rail Transport

Rail is a useful method of transportation, especially if the animals can be delivered to the railway station shortly before the train is due to leave, and collected without delay on arrival at their destination. It is relatively cheap, but some exposure of the animals on the platform while waiting for the train cannot be avoided, and they may suffer for a short time from heat or cold. If changes of train are necessary, this exposure will be repeated at each stage.

3. Air Transport

For longer distances air transport works well, and many airlines are very experienced in handling live animals. As far as possible direct flights should be chosen. Arrangements should be made for looking after the animals if the flight out is delayed or cancelled, and for collecting them at the destination. Long waits in warehouses, either at the airport of embarkation, on arrival at the destination, or at any intermediate stops, must be avoided.

Several international airlines have prepared cargo manuals in which are included detailed recommendations for the packing and carriage of many species of animals. There may also be local or national regulations controlling this traffic. It is not possible to give full details in a book of this kind because changes in detailed requirements take place from time to time. The most useful source of information about conditions of air transport is usually the airline that is to carry the animals.

The airline will also be able to give information on documentation, which is sometimes quite extensive; and on import or export regulations, which vary considerably in different countries and for different species.

At a few major airports there are animal hostels for the temporary accommodation of animals, either in transit or awaiting despatch or collection on arrival. Such hostels provide a valuable service, and their charges are usually very reasonable. There is a very good one at London Airport (Heathrow) run by the Royal Society for the Prevention of Cruelty to Animals.

D. TRAVELLING CONTAINERS

Various bodies have made recommendations about the conditions in which laboratory animals are carried by air. The British Standards Institution has published a number of standard recommendations, under the code B.S. 3149. There are also recommendations of the Institute of Laboratory Animal Resources (USA), to which reference may be made.

Travelling containers must, first of all, be escape-proof. One of us once spent an unhappy half hour in the body of a large animal freight plane, in flight over Switzerland, catching nine rhesus monkeys that had escaped from a damaged container. But even an escaped mouse can be a serious cause of concern in an aircraft, and no pilot will take off if he thinks there may be an animal of any kind loose in his plane.

Containers must be light, consistent with strength, and well ventilated, so that stacking them in the confinement of an aircraft cannot lead to suffocation. They must also be easy to handle, so that loaders will neither drop them nor risk getting their fingers bitten by the animals inside. They must carry prominent labels.

The correct method of labelling containers of laboratory animals is vitally important, and once again the carrier is a useful source of advice. When animals are being carried across international boundaries, it should

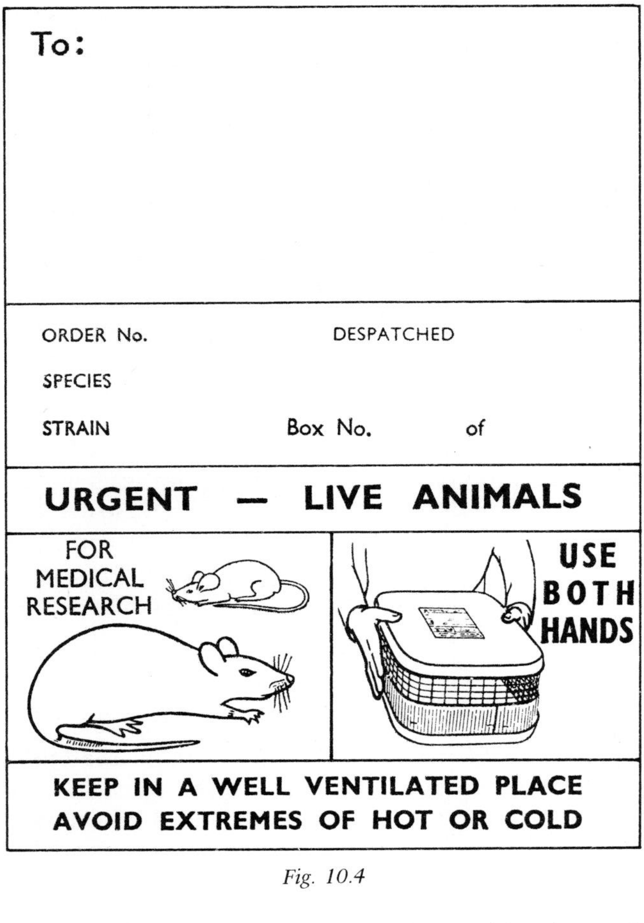

To:

ORDER No.   DESPATCHED

SPECIES

STRAIN   Box No.   of

URGENT — LIVE ANIMALS

FOR MEDICAL RESEARCH

USE BOTH HANDS

KEEP IN A WELL VENTILATED PLACE
AVOID EXTREMES OF HOT OR COLD

Fig. 10.4

be remembered that the containers will be handled by those who may be ignorant of the language of the country of despatch. Labels should, therefore, carry outline pictures indicating that there are live animals in the containers and that therefore they should be handled gently, not dropped or crushed, and not left exposed to blazing sun, rain or cold wind (Fig. 10.4). A special care arises with animals being sent in filter cages (such as germfree animals) from one country to another. If the cages are opened in transit, for customs inspection or for any other purpose, the value of the animals will be totally destroyed. The International Air Transport Association (IATA) recommend that a special label be used in such cases, and this is illustrated in Fig. 10.5.

Several types of travelling containers are illustrated in the British Standards already referred to; and in Short and Woodnott (1969) and

Laboratory Animals

These animals are in a

FILTER CONTAINER

to exclude germs.

If it is necessary to inspect them
for any purpose, this must only
be done under the direction of the
consignee. If they are otherwise
opened |, or if they are given
food or water, their value will
be totally destroyed.

DO NOT OPEN, FEED OR WATER

Fig. 10.5

UFAW Handbook (1971). A metal and plastic container for rats and mice is illustrated in Fig. 10.4. Large animals may be carried in any of the recommended containers, made of wood, metal or other materials. A convenient container for, say, a cat may be made from a plastic box or bin with an airtight cover; holes or windows are cut in the plastic for ventilation, and covered either with wire mesh or with a filter material which will protect the animal in transit from the danger of picking up infection.

Laboratory animals are valuable products. When they are sent from one place to another, proper thought must be given to the package containing them. It is not unreasonable that a travelling container should cost 5–6 per cent of the value of the animals inside it; indeed, by normal packaging standards this is a low figure. In most cases it is enough to provide a safe, comfortable and satisfactory container.

E. DESPATCH AND RECEPTION

No trouble is too much to ensure that the right animals are packed, the container correctly labelled, and the animals sent at the right time and by the correct route. This may seem elementary, but when many consignments are going out together, mistakes can easily happen. Therefore any system of despatch should incorporate double checks at every stage, from final weighing and inspection to counting, packing and loading.

At the destination the animals should be received by a responsible person: a veterinary surgeon or biologist on the staff of the laboratory, or a senior animal technician. They should be inspected and settled down in their new quarters, given food and water, and any sick or injured animals noted and reported to the breeding unit. In no circumstances should they be left on an unloading dock until a storeman can find time to deal with them.

Experience has shown that, if reception is left to junior or inexperienced staff, mistakes such as mixing of sexes or weight/age groups are liable to happen. When they come to light some days later, recriminations are inevitable.

In summary, there are no great difficulties in sending laboratory animals even quite considerable distances from the breeding unit to the user's laboratory, provided simple precautions, which are mainly common sense, are observed.

CHAPTER 11

Organization

I. Introduction

Biological work entailing the use of laboratory animals is, like all other scientific activities, world-wide in scope. It follows that the provision of laboratory animals must be similarly considered at an international level. The production, care and use of laboratory animals demands some knowledge of many biological disciplines, and although it is perhaps a little presumptuous to talk about laboratory animal science, there is no doubt that the problems encountered in this field have to be considered scientifically.

It is not surprising, therefore, to find that there are scientists whose whole professional interest is in laboratory animal matters, and that there has come into existence a number of professional bodies concerned with this subject. By the same token, the subject is international in interest, and there is much exchange of information and animals across national boundaries.

The use of animals in the laboratory has led to a certain amount of disquiet in the minds of many reasonable people, who have imagined that perhaps on occasion animals in the laboratory are subjected to experimental procedures that are painful, even to the limits of cruelty. In this, such people have been urged to believe the worst of scientists by a dedicated, if misguided, minority of anti-vivisectionists, those enemies of knowledge as A. V. Hill (1929) called them, who deny both the utility and the morality of animal experiments. The perfectly reasonable disquiet, and the unreasonable attacks on all users of experimental animals, have led to legislation designed to protect the animal from possible abuses, and to regulate what many people think should in any event be regulated. For there are ethical considerations here, and it is better to bring them out into the open than to pretend they do not exist or are irrelevant.

This chapter will cover all these aspects of the laboratory animal field.

II. Laboratory Animal Centres

In 1945 an unofficial committee produced a report on the supply of laboratory animals in Great Britain. This committee had been set up some 3 years previously by a number of scientific societies whose members used animals in the course of their work and were experiencing difficulties, exacerbated by war conditions, in obtaining them. The unofficial committee had attempted the first survey of the production and utilization of laboratory animals ever to be undertaken and published (although the publication of its findings was restricted and is now out of date), and it made certain recommendations. Among them was, first, that an advisory committee be set up, on a national basis, to examine the whole problem in greater detail and, second, that the work thereby entailed should be delegated to a laboratory animals bureau with a whole-time scientific director and suitable staff. Eighteen months later, in the spring of 1947, the Medical Research Council, acting on these recommendations, set up an advisory committee and created a new unit on the lines suggested. Today that unit is known as the Laboratory Animals Centre.

A. THE LABORATORY ANIMALS CENTRE (UK)

A detailed account of the work of the British Laboratory Animals Centre will be found in the UFAW Handbook (1967). It is only necessary here to summarize its more important activities.

The Laboratory Animals Centre does not aim to supply animals directly for experimental use (unlike centres in some other countries). It does, however, maintain a number of primary type colonies of several species, especially of inbred strains of mice, and from them it supplies breeding nuclei for expansion elsewhere. It also maintains colonies in gnotobiotic conditions and in conditions of rigid disease control, and from these can assist other laboratories to found new healthy colonies of their own.

The Centre provides a wide advisory and information service, both to laboratories using animals and to breeders. In particular, the Centre organized, in 1950, a scheme for the accreditation of commercial breeders of laboratory animals. For several years this scheme was confined to mice, guinea-pigs and rabbits, but recently it has expanded to include many other species.

The accreditation scheme aims to set minimum standards for animals produced by breeders for sale to laboratories. These standards have always attempted to be realistic; that is, they are high enough to be a useful guide to the laboratory, but not so high that most commercial breeders would be unable or unwilling to meet them. As knowledge advances, and techniques in this field improve, so the standards rise; the accredited animal today is greatly superior to its ancestor of 20 years ago.

In conjunction with the accreditation scheme, the Laboratory Animals Centre publishes twice monthly lists of animals currently available from commercial breeders. These lists are called *Parade State*, and are a useful purchasing guide.

The Centre publishes, every January and July, a *News Letter*, giving current news of its activities. It also publishes, in conjunction with ICLA, (p. 251) *Mouse News Letter*. Other publications include an *International Index of Laboratory Animals* (also in conjunction with ICLA); a series of Manuals; and *Guinea-pig News Letter*. Full information about all of them can be obtained from the Laboratory Animals Centre, Medical Research Council Laboratories, Woodmansterne Road, Carshalton, Surrey.

Not least important is the research programme carried out at the Centre, which has greatly expanded in recent years. It is a healthy sign that opportunities exist at such a place to pursue lines of investigation that are known to be of practical, perhaps urgent, importance. Indeed, from its inception the Centre (originally called the Laboratory Animals Bureau) has recognized the importance of including relevant research among its many activities, and its staff have made significant contributions to our knowledge of laboratory animals, and will continue to do so. One particularly important field, where the Centre has unique opportunities for advancing knowledge of practical value, is in the setting of standards for commercially bred laboratory animals, already referred to. Ideas of the exact nature of these standards, and how they are to be achieved, as well as what the laboratories really need, are constantly evolving: many of them have been discussed elsewhere in this book, and laboratory animal centres, wherever they are, have a duty to throw whatever light they can on this difficult problem.

B. OTHER NATIONAL CENTRES

Today there are laboratory animal centres in many other countries throughout the world. They are listed as follows, in alphabetical order of countries:

Australia	Laboratory Animal Centre of Australia, Institute of Medical and Veterinary Science, P.O. Box 14, Rundle St, Adelaide, South Australia.
Czechoslovakia	Information should be sought from the ICLA National Member, Dr M. Šeda, Research Institute for Pharmacy and Biochemistry, 17 Kouřimská, Praha 3.
France	Centre de Sélection et d'Élevage des Animaux de Laboratoire, 45 Orléans-La-Source.
Germany *(Federal Republic)*	Zentralinstitut für Versuchstierzucht, 3 Hannover-Linden, Lettow-Vorbeck-Allee 57, Postfach 20345.
Hungary	Laboratory Animals Institute, Táncsics M. Road, Gödöllő.
India	Laboratory Animals Information Service, Cancer Research Institute, Parel, Bombay-12.
Netherlands	Centraal Proefdierenbedrijf TNO, Zeist. Woudenbergseweg 25.
Poland	Laboratory Animals Centre of the Polish Academy of Sciences, Łomna-Las, Post Czosnów/Nowy dwór Maz.
South Africa	National Committee on Laboratory Animals, c/o South African Medical Research Council, Scientia, Pretoria (Private Bag 380).

Union of Soviet Socialist Republics	Laboratory of Experimental Animals, c/o Academy of Medical Sciences, Solyanka Street 14, Moscow.
United Kingdom	Laboratory Animals Centre, Medical Research Council Laboratories, Woodmansterne Road, Carshalton, Surrey.
United States of America	Institute of Laboratory Animal Resources, 2101 Constitution Avenue, Washington, D.C. 20418.

C. OUTLINE OF FUNCTIONS

The functions of a laboratory animals centre are four in number: the collection and dissemination of information, the maintenance of primary type colonies, the conduct of relevant research, and the training of personnel.

1. Collection and Dissemination of Information

This first need is to obtain some reasonably accurate information about the problems that the centre has to undertake. This demands a survey of the production and utilization of laboratory animals throughout the whole country. It cannot be too strongly emphasized that until such a survey has been carried out the centre will have no reliable guide to the direction in which it should bend its efforts. Moreover, the carrying out of a survey will bring the existence of the centre to the notice of those it aims to help, and so lay the foundations for a continual feed-back of information from the users to the centre.

To serve its purposes, a centre must be widely known, and known to be useful. It must strive to become the first place to which laboratory workers look for the answers to any questions that arise in their use of, or need for, animals.

Only in this way can a balanced picture be built up of topical problems, and of knowledge of their successful solution in some other laboratory.

Current shortages and surpluses of different species and strains of animals should receive careful consideration, so that as far as possible they may be balanced as between different laboratories. The frequent publication of a list of animals currently available can often be useful.

2. Primary Type Colonies

A laboratory animals centre is the obvious place for setting up and developing a centre for primary type colonies. As has been pointed out in previous chapters, the control of such colonies requires a considerable expenditure of skill and time on the part of the scientific staff of the centre. They are for this reason best fitted to provide the expert advisory service which the centre aims to offer to other laboratories. It need hardly be added that those who presume to offer such advice must be constantly in practical touch with the problems on which they are advising, and the maintenance of primary type colonies provides just the sort of contact and experience that is needed.

It is possible that certain strains, for which there is a very restricted demand outside one or two specialized fields, are already maintained in laboratories specializing in those fields. This applies particularly to some strains of mice or rats used in cancer research laboratories. In such cases it would often be not worth while for the centre to duplicate colonies of these strains, but rather to ask the laboratories already maintaining them to regard them as primary type colonies and be prepared to furnish occasional breeding nuclei—white-label stocks—to other workers. There is no reason why the centre, with the agreement of the other laboratory, should not include these strains in its catalogue of primary type colonies, provided it is satisfied that the conditions that merit classification as a primary type colony are being met by the other laboratory. The degree to which outlying primary type colonies and those maintained in the centre are included under one system will depend on the resources of the centre and the opportunities presented by other laboratories. In the early life of a national centre almost the whole system may be external to the centre, but as the centre acquires laboratories and animal houses, some at least of the colonies will be maintained internally.

The production of large numbers of animals for experimental use is not a necessary function of a laboratory animals centre, but there is no reason why it should not be undertaken as an additional activity. Production units must of course be physically separated from the primary type colony centre. A production unit will normally have to run on economic lines, if it is supplying animals to other laboratories; it may even make money, although this is not its main purpose. Unless it is to charge its customers a price for animals that will appear to them unduly high, the primary type colony centre itself cannot expect to be self-supporting, but the value of a laboratory animals centre cannot be measured by its immediate income and expenditure account. It has to be assessed by the effect it has on the usefulness of animals in every laboratory in the country.

3. Research

Current problems require current research to solve them. The scientific staff of a centre have the job of controlling the quality of their primary type colonies and of providing help and guidance to other laboratories. If they are competent to do this, they will be of a calibre that will lead them to seek opportunities for research, for which they must, as a necessary qualification, also be competent. Fortunately, in this context at all events, there is no lack of relevant research problems regarding laboratory animals.

The main fields of research will naturally be genetics, hygiene and nutrition, the three legs of a tripod on which the quality of laboratory animals rests. This points immediately to the need for three members of the scientific staff of the centre, specializing in these disciplines. Husbandry, which has something to do with both hygiene and nutrition, but also includes care and control of the physical environment, may be the main interest of a further investigator. Already the centre is a considerable scientific group, but this only indicates that the control of quality must be a continuing preoccupation of the centre, and not a pious hope devoid of any means of realization. If good animals are really needed, then sufficient scientific resources must be devoted to meeting this need. No animals were ever produced by committees or resolutions.

4. Training

A laboratory animals centre offers certain advantages for training both scientific and technical personnel. The sort of scientists that are likely to seek special training in this field are those who intend to become curators of animal divisions elsewhere. They will need to know something about the special techniques of inbreeding and other genetical manipulations; of hygiene and the control of health, including the early recognition, diagnosis and control of ill-health; of nutrition; and of the countless details of husbandry that have to be watched. A period of training in an established centre can give them much of what they will need when they come to take charge of an animal division.

Whether a laboratory animals centre should be a training school for animal technicians is a point on which more than one opinion could be held. Certainly the centre's own animal technicians will need to undergo formal and rigorous training in both the general and the special techniques that are practised in the centre. In certain circumstances student technicians from elsewhere could be satisfactorily trained in the centre, with emphasis on teaching them the skills that are common to all animal houses. But in many respects the training of animal technicians is better given in the places in which they normally work, reinforced by a regular programme of lectures and demonstrations and, perhaps, by short intensive courses in other laboratories that can show them species and methods not used in their own animal house.

III International Committee on Laboratory Animals (ICLA)

The international character of the provision and use of laboratory animals gained a degree of formal recognition in 1956, by the setting up of an International Committee on Laboratory Animals (ICLA). The history of this event, and the subsequent development of ICLA, have been recorded in *ICLA Bulletin* **26** (1970). In the same bulletin is published the most recent constitution of ICLA. The bulletin is published twice yearly, in March and September, and in each issue is a directory of members. In March 1971 there were National Members of ICLA in Argentine, Australia, Belgium, Bulgaria, Canada, Czechoslovakia, Denmark, France, German Democratic Republic, Federal Republic of Germany, Greece, Hungary, Iceland, India, Ireland, Israel, Italy, Japan, Lebanon, Netherlands, Nigeria, Norway, Poland, Romania, South Africa, Spain, Sweden, Switzerland, Turkey, United Kingdom, United States of America, and Union of Soviet Socialist Republics.

The Governing Board consisted of representatives of the Council for International Organizations of Medical Sciences (CIOMS), the International Union Against Cancer (UICC), the International Union of Biochemistry (IUB), the International Union of Biological Sciences (IUBS), and the International Union of Physiological Science (IUPS), together with five other members, including the secretary, who is National Member for Norway, Dr S. Erichsen, National Institute for Public Health, Geitmyrsveien 75, Oslo 1, Norway; and the treasurer, who is National Member for France, Dr M. Sabourdy. Centre de Sélection et d'Élevage des Animaux de Laboratoire, 45 Orléans-la-Source, France.

On the general committee there were also five Associate Members and one Individual Member.

Under the 1969 Constitution, the Governing Board of ICLA meets twice a year, and all authority of the organization is vested in it from one General Assembly to the next. General Assemblies, comprising the whole membership, are held about once in 3 years: the last was held in France in 1969.

Up to 1962 ICLA received most of its financial support from the United Nations Educational, Scientific and Cultural Organization (UNESCO); together with some small but important contributions from the international unions supporting it. Subsequently, ICLA has had annual contracts with the World Health Organization (WHO), which has provided much-needed funds to enable it to carry out its work. The unions have also continued their support, and some contributions have also come from Associate Members and from other sources.

In its early years ICLA helped to organize a number of national surveys on the supply, quality and use of laboratory animals, and these were subsequently published. Surveys were carried out in India, Italy, Japan, Switzerland and United Kingdom (published in 1958): in Denmark, Finland, Iceland, Norway, Sweden and Turkey; in Western Germany and Austria; and in Australia (published in 1959): in France, Israel and Western Germany (supplementary survey); in Belgium, Czechoslovakia, Luxembourg, Netherlands, Poland, South Australia (supplementary survey), and

United States of America (partial survey) (published in 1962): and in Argentine, Brazil and Chile (published in 1964). These surveys brought to light, for the first time, the quantitative as well as the qualitative problems of the supply and use of laboratory animals, and they have been an essential foundation, for all their incompletenesses, on which a rational laboratory animal policy can be built.

In 1958, less than 2 years after its foundation, ICLA organized an international symposium on *Living Animal Material for Biological Research*. The proceedings were published by ICLA. A second symposium, on *Problems of Laboratory Animal Disease*, was held in 1961, and the proceedings published in 1962. The third ICLA symposium, on the *Husbandry of Laboratory Animals*, was held in 1965, and the proceedings were published in 1967; and the fourth, held jointly with the Institute of Laboratory Animal Resources (USA), was held in 1969, on the subject of *Defining the Laboratory Animal in the Search of Health*, and the proceedings published in 1971. (The full references of all these publications will be found in the Bibliography.)

In addition to these more classic activities, ICLA has engaged in a great variety of other activities. It has provided scholarships for scientists and technicians to visit foreign centres and learn special techniques, and it has sent experts to various countries to provide specialist help and advice. It has listed definitions of terms commonly used in connection with laboratory animals, published a cumulative bibliography (now discontinued), promoted the training of animal technicians, and set up working parties for a number of technical purposes. And its work is continuing and growing.

Above all, ICLA is an international point of focus for all who have a special interest in laboratory animals, and wish to inform themselves of current activities in countries other than their own. The *ICLA Bulletin* has a wide circulation, and carries national news items, as well as much else. There is little doubt that ICLA has stimulated interest in laboratory animals in several countries, and that it will do so yet in others in the future.

IV. Professional Organizations

In addition to laboratory animal centres, which are national bodies, and to ICLA, which is international, there are several professional, scientific and technical organizations concerned with laboratory animals. They are, of necessity, national in character, but not exclusively so. Most of them admit foreign members, provided they qualify for membership. These qualifications vary considerably, and the following comments are not intended to be definitive; for full details, information should be sought from the organizations themselves, whose addresses may be obtained from their national representative on ICLA (see current *ICLA Bulletin*).

A. NATIONAL SOCIETIES

Australia	Animal Technicians Association.
Canada	Canadian Council for Animal Care.
Denmark	Scandinavian Federation for Laboratory Animal Science.
France	Association des Techniciens d'Animaux de Laboratoire.
Germany	Gesellschaft für Versuchstierkunde.
Japan	Japan Experimental Animals Research Association.
Netherlands	1. Animal Technicians Association.
	2. The Netherlands Society for Laboratory Animal Science.
Norway	See *Denmark.*
Sweden	See *Denmark.*
United Kingdom	1. British Laboratory Animals Veterinary Association.
	2. Institute of Animal Technicians.
	3. Laboratory Animal Breeders Association.
	4. Laboratory Animal Science Association.
	5. Research Defence Society.
	6. Universities Federation for Animal Welfare.
United States of America	1. American Association for Laboratory Animal Science.
	2. American College of Laboratory Animal Medicine.
	3. Animal Welfare Institute.
	4. Laboratory Animal Breeders Association.
	5. National Society for Medical Research. In addition, there are several state societies and associations in USA.

B. FUNCTIONS

The functions of these professional organizations vary considerably, partly according to their membership, and partly in the light of their corporate aims. Some are mainly concerned with the training and qualifications of animal technicians. Some concentrate on providing specialist training and recognition for veterinarians. Some attract members from many disciplines who want to exchange knowledge about laboratory animals, without necessarily seeking any professional advancement thereby. Some represent the interests of commercial breeders. Some are mainly concerned with countering the attacks on research by anti-vivisectionists; while others have as their main interest the welfare of laboratory animals consistent with the interests of research.

All these organizations, with the exception of the Research Defence Society and the Universities Federation for Animal Welfare, have come into existence from 1950 onwards. They are striking evidence of the explosion of professional interest in laboratory animals during the last 20 years. Most of them organize scientific meetings, conferences and symposia at regular intervals, usually once or twice a year; and some of them sponsor their own professional or technical periodicals, reference to which is made in the general bibliography.

Qualification for membership varies. Two or three are exclusively for veterinarians, others are more liberal, admitting to membership scientists, animal and laboratory technicians, and often those without any formal qualifications but having an interest in the objects of the organizations. Many people whose work is mainly with laboratory animals will belong to several of these professional bodies.

The appearance in recent years of so many new professional organizations concerned with laboratory animals—and the foregoing list is certain to be incomplete when these words are read—indicates a tremendous development of interest in the subject. Little gift of prophecy is needed to predict that, with so much intellectual and practical effort being expended, the way laboratory animals are produced and used, and even the nature of the animals themselves, will undergo profound changes; and this is indeed already happening. The health status that is taken for granted today was, less than 20 years ago, thought of as a possibly unattainable ideal, even a pipe dream. Developments in progress today are likely to produce equally radical transformations in the coming years.

One particularly important function of laboratory animal organizations is that of special training and qualifications at the professional and technical levels. The British Institute of Animal Technicians (formerly the Animal Technicians Association) was the first to introduce certificates and diplomas, obtained by examination, and their qualifications have gained wide recognition (see Chapter 8). The American College of Laboratory Animal Medicine has for many years offered a diploma, obtainable by examination, to veterinarians in this field. Other qualifications are being considered elsewhere.

V. Legal Considerations

A. LEGAL CONTROL OF EXPERIMENTS

In 1875, as a result of some public agitation which raised echoes in the Palace of Westminster, a Royal Commission on the Practice of Subjecting Live Animals to Experiments for Scientific Purposes was appointed, and in the following year, as a result of its recommendations, the Cruelty to Animals Act was passed in Parliament. This was the first law to be introduced anywhere to control the use of animals for scientific enquiry. A second Royal Commission was appointed in 1906 and reported some 6 years later, and some of its recommendations were subsequently adopted. In 1963 a Departmental Committee was appointed, under the chairmanship of the late Sir Sydney Littlewood, and reported in 1965 (Report, 1965). Its recommendations are still under consideration by the government.

In fact, the Act of 1876 still controls animal experimentation in Great Britain. It is in some senses a triumph of adaptation that makes this control, after nearly a century in a field that has developed so colossally, work at all, but it does work astonishingly well in spite of its age.

An outline of its operation will be found in *Notes on the Law Relating to Experimental Animals in Great Britain* (see Bibliography). Briefly, experiments on living vertebrate animals that are likely to cause pain must be carried out in places registered by the Home Office, by persons holding licences authorizing them to perform such experiments.

The registration of the place enables the Home Office to be aware of, and inspect, all places where such experiments may take place. The licence is personal to the holder, and its scope may be modified by one or more certificates, as laid down in somewhat unimaginative detail in the Act.

Places are subject to periodical inspection, not less than once a year, many of them much more often. The inspection is no mere formality: on the contrary, it is taken very seriously both by the inspectors, who all possess medical or veterinary qualifications, and by the licensees. An annual report on the working of the Act is prepared by the chief inspector, and placed before Parliament.

The report of the Littlewood Committee (Report, 1965) is lengthy—a total of 256 pages—and its specific recommendations number eighty-three. As in the case of the reports of the two Royal Commissions, it found no evidence of abuse of animals by scientists, but it followed their examples by proposing further restrictions on the way scientists should use animals, some of which would require new legislation.

However, most British scientists find they can work well within the restriction of the Act of 1876, and if they do occasionally find it interferes with their work, they tend to take the view that it is a small price to pay for the observance of a proper restraint in the use of animals and of a proper respect for public opinion. Very few would like to see the Act of 1876 repealed.

There are laws in several other countries that exert more or less control over the use of animals in the laboratory, and in general they seem not to hinder research, while at the same time they afford some protection for the animal. Some countries have no legal control at all, and there abuses are possible. Most humane people feel that, because the opportunities exist in the laboratory, as much as they do in many other places where animals are used by man, of inflicting severe suffering on animals, a measure of formal restriction and protection is advisable, if only to protect the good name of the scientist. Many scientists would say that such control is not just advisable but necessary; cruelty, the wanton infliction of pain without sufficient reason, is a human failing, and it has to be expected that scientists will not be totally immune from this or other such failings (although they are certainly not, as a class, the sadists that some anti-vivisectionists make them out to be). They are in all probability rather less likely to be cruel or insensitive to others' pain than the generality of people, but on the whole the above sort of legislation can be valuable to them and to the animals.

B. PROCUREMENT OF STRAY ANIMALS

In most large urban centres throughout the world there is a quasi-wild population of dogs and cats, which have no owner and no settled home. They may have been domestic pets at one time, but they have been rejected, or lost, and have learned to live on the pickings of human habitations, and on the vermin that may infest them.

Cities have the task of keeping this population of strays within tolerable bounds. This entails catching the strays and, if no home exists or can be found for them, killing them. It is a task which often falls to the animal welfare societies, which rightly take the view that a painless death is preferable to a squalid and precarious life of disease and starvation. Most of these societies also take the view that to offer stray dogs and cats for laboratory use is not an acceptable alternative to a painless death on their own premises.

In Great Britain the law requires stray dogs to be handed over to the police, and it prohibits the police from giving or selling the dogs for "vivisection". In practice, they are handed over to one of the humane societies for destruction; and for the most part the same fate is visited on stray cats, although the law does not insist as it does for dogs. Many times the number of stray dogs and cats are thus destroyed as would be required for use in laboratories; but it should be added that for many, if not most, laboratory purposes a stray animal is not suitable material. It is, however, cheap, and for dissection, teaching and some non-recovery experiments the low cost would outweigh the poor quality. But this source is almost unavailable in Great Britain.

It is, however, an important source of laboratory dogs and cats in many other countries. In the United States of America laws have been passed in many states, making stray dogs and cats available to laboratories, subject to safeguards about ownership, defraying expenses and so on. Many animal welfare societies have in recent years fought these so-called pound laws, but with little success. In 1969 more than 100,000 dogs and 30,000 cats were obtained from American pounds. Great bitterness has been generated in this dispute, and it is fair to ask whether the saving of money on the one hand, made possible by using pound animals, is enough to counterbalance, on the other hand, their poorer quality as compared with specially bred dogs and cats and the ill-will against the scientist aroused by their use.

C. IMPORT AND EXPORT REGULATIONS

Almost all countries have laws and regulations governing some aspects of the trade in animals, including laboratory animals. These regulations vary from the total prohibition of importation of certain species, through importation with statutory quarantine, to free import. Export of certain species may also be prohibited or controlled.

Carriers, such as airlines, nearly always impose further conditions of acceptance of live animals as freight.

The regulations are sometimes complicated, and they are changed from time to time. Useful sources of detailed and up-to-date information are the appropriate government departments in UK, that is, those concerned with health, agriculture and trade; the scientific or commercial attaché of the embassy of the foreign country with which it is proposed to trade; and the airlines who may carry the animals. They will be able to give information about veterinary certificates as well as import and export licences, and any other documentation or restriction.

The conditions of transportation are discussed in Chapter 10.

VI. Public Relations

The legal control of animal experimentation has been introduced in Great Britain, and in several other countries, as a result of the pressure of public opinion. It is not, perhaps, an issue of the first importance, but it does arouse deep emotions, and it is well for the scientist to ponder carefully about his public relations in regard to animal experiments; for his business is not his alone, but society's as a whole. He needs the general confidence of the society in which he lives, and he will not receive it if he fails to ensure that intelligent laymen can, with reasonable facility, find out just what he does with his animals and why.

A. 'ENEMIES OF KNOWLEDGE' (Hill, 1929)

Animal experimentation has generally possessed a bad public image, cultivated sedulously if misleadingly by the anti-vivisectionists, who would prohibit it altogether. But medical science has proceeded despite this, although it has been embarrassed at times, and occasionally hindered by it. There are anti-vivisection groups in most countries where there is animal experimentation, and the march of medical progress impresses them not at all. Most anti-vivisectionists are emotionally entangled in their cause to an extent that makes them purblind, however rational they may be on other subjects. Psychologists will no doubt have ingenious explanations for this phenomenon, some not too flattering to those affected by it, but it is not something that is amenable to the voice of logical argument, nor can it be brushed aside as a trivial foible. Although there are some anti-vivisectionists who delight less in demonstrable truth than in boffin-bashing and lurid imagery, there are many who are intelligent and sincere, and who feel very deeply that pain ought not to be inflicted on animals for any reason. It is those to whom the scientist must address himself: he may never bring them round to his point of view, but he may persuade them that he is entitled to hold it himself.

There is also the mass of normal, more or less intelligent people who have seldom given a thought to animal experiments. They trust the scientist to behave as they would want him to: or they would trust him if they ever thought about it. But, equally, if they are told that scientists subject animals to unspeakable cruelties, excusing themselves unjustifiably in the name of progress, and they are totally ignorant of the facts, they can hardly be blamed for accepting a *prima facie* case against science. It is here that public relations can do the most good.

Experience has shown, over and over again, that to present the facts honestly and without constraint is the surest way to gain the confidence of an indifferent public. Openness, in this or in many other subjects, is a shield which the shafts of anti-vivisectionists' propaganda cannot penetrate.

Anti-vivisection societies—and there are many of them—attract financial support, including occasional substantial legacies, and they can spend all their efforts in denigrating experiments on animals. The scientist only has his spare time to reply. But despite the advantage of money and of time, the anti-vivisectionists' cause is unlikely to prosper, for they lack the most powerful weapon of all, truth.

B. ANIMAL WELFARE SOCIETIES

Apart from the anti-vivisection societies, which really have little or no positive effect on animal welfare, there is in most countries a number of animal welfare or humane societies, whose objects are the welfare of animals and their protection against cruelty. Most of such societies understand the need for animal experiments, and more or less reluctantly tolerate them. Among their membership there are bound to be some vigorous opponents of animal experiments, but there will also be others whose sympathies are as much with the scientist as with the animal, and see there is no insoluble problem.

These genuine animal welfare societies are a credit to any civilized community, and if they occasionally show less than total support for experimental science this should not be taken amiss. Moreover, when it comes to practical considerations it will often be found that a society that has publicly leaned too far towards the anti-vivisectionists' cause will at the same time provide valuable help in, for example, the care of laboratory animals in transit from one place to another. It is no accident that the animal hostel at London Airport is run by the Royal Society for the Prevention of Cruelty to Animals.

It is also no accident that the first text-book on laboratory animals was sponsored by the Universities Federation for Animal Welfare (UFAW, 230 High Street, Potters Bar, Hertfordshire), thus demonstrating that animal experimentation and a concern for the welfare of laboratory animals were not incompatible. *The UFAW Handbook on the Care and Management of Laboratory Animals* is now in its fourth edition, and has achieved world-wide acclaim. It is referred to frequently in this book. An organization with aims similar to that of UFAW is the Animal Welfare Institute (USA). It publishes some useful material on laboratory animals, which is referred to in the Bibliography (see *Basic Care of Experimental Animals* and *Comfortable Quarters for Laboratory Animals*).

C. THE RESEARCH DEFENCE SOCIETY

At the beginning of this century anti-vivisectionist activity was powerful and virulent, and was largely responsible for persuading the government of the day to appoint a second Royal Commission in 1906. In 1908 a group of scientists who had given evidence to the Royal Commission decided that, no matter how the commission reported, the anti-vivisectionists were certain to continue their activities, and therefore that the scientists should form a society to counteract this mischief. Under the inspiration of Stephen Paget, a surgeon and a poet, they founded the Research Defence Society, "to make known the facts about experimental research involving the use of animals, and the conditions and regulations under which animal experiments are conducted in the United Kingdom; to emphasize the importance of such experiments to the welfare of mankind and animals and the great saving of human and animal life and health and the prevention of suffering already due to them; to defend research workers in the medical, veterinary and biological sciences against attacks by anti-vivisectionists; and to help workers in drawing up their applications to the Home Secretary for the licence and certificates needed for the proper conduct of experiments on animals".

The Research Defence Society (11 Chandos Street, Cavendish Square, London W1) has had a distinguished career, and a brief account of it may be read in a pamphlet published by the society. Today it is still active in the cause of the experimental biologist, and it helps to keep the anti-vivisectionists from having the argument all their own way. It publishes an annual journal *Conquest*, which includes the annual Stephen Paget Memorial Lecture; and it has published a series of most useful *Conquest Pamphlets*, which explain in simple language the debt that human and animal life owes to animal experimentation. It sponsored the publication of a substantial book on the same subject (Lapage, 1960).

The Research Defence Society has good relations with all British Universities, and with the Home Office, who administer the Act of 1876. All who hold licences under that Act should regard it as an obligation to belong to the Society.

An American Society with aims similar to that of the Research Defence Society is the National Society for Medical Research.

VII. Conclusion

In his presidential address to the British Association, which met in Cardiff in 1960, Sir George Thomson said:

". . . The best way to make advances in technology, whether on the medical or engineering side, turns out to be to understand the principles. This is quite a recent discovery—indeed it has probably only recently become true. It would not have been much use, for example, to man in the stone age, or even a few hundred years ago, to try to understand the principles of tanning with no basic knowledge of chemistry to guide him. He did better by trial and error. Even the steam engine was developed with little knowledge of the determining principles, though the best scientific minds of the day were much interested, and the thought they gave it advanced science by discovering thermodynamics."

If this argument is true, it explains why a practical guide, which is what this book has attempted to be, has had to concern itself so much with principles. In the matter of laboratory animals we have emerged from the stone age period of trial and error. We find, to the surprise of some, perhaps, but to the satisfaction of most of us, that it is a scientific subject of increasing importance. The ideal of an animal comparable in purity and definition with a chemical reagent is far off and even unattainable. This is, however, no excuse for continuing to accept for biological investigation an unknown conglomeration of genes, parasites and morbid affections, contained in a vehicle that is as much an offence to the eye as to the nose.

The techniques of producing good laboratory animals are fairly well understood, although many problems still remain unsolved. Their first cost will be higher than formerly, but their ultimate cost, in terms of their usefulness in research, is likely to show a handsome dividend on investment. We know enough today to be safe in turning our back on the stone age.

References

Bailey, D. W. and Usama, B. (1960). A rapid method of grafting skin on tails of mice. *Transplantn Bull.* 7, 424.

Barber, B. R. (1970). Development of a one-piece cap for animal drinking bottles. *Bull. Inst. Anim. Tech.* 6, 8.

Bennett, N, K. (1969). The preparation of small quantities of pelleted diets for experimental purposes. *J. Inst. Anim. Tech.* 20, 57.

Beveridge, W. I. B. (1960). Economics of animal health, *Vet. Rec.* 72, 810.

Brown, A. M. (1963). Skin grafting of mice using the Waldemar type punch. *J. Anim. Tech. Assn* 14, 11.

Charles, R. T., Stevenson, D. E. and Walker, A. I. T. (1965). The sterilization of laboratory animal diet by ethylene oxide. *Lab. Anim. Care* 15, 321.

Charles, R. T. and Walker, A. I. T. (1964). The use of ethylene oxide for the sterilization of laboratory animal foodstuffs and bedding. *J. Inst. Anim. Tech.* 15, 44.

Cholnoky, E., Fischer, J. and Józsa, A. (1969). Aspects of genetically defined populations in toxicity testing. 1. A comparative survey of populations obtained by different breeding systems and "mosaic populations". *Z. Versuchstierkunde* 11, 298.

Coates, M. E. (1970). The sterilization of laboratory animal diets. In *Nutrition and Disease in Experimental Animals*, p. 38 (see Bibliography).

Coates, M. E., Ford, J. E., Gregory, M. E. and Thompson, S. Y. (1969). Effects of gamma-irradiation on the vitamin content of diets for laboratory animals. *Lab. Anim.* 3, 39.

Cobb, K. W. (1963). Lighting criteria for animal housing. *Lab. Anim. Care* 13, 332.

Cohen, B. J. (1960). Organization and functions of a medical school animal facility. *J. Med. Educ.* 35, 24.

Coid, C. R. (1968). Building design in relation to function of a laboratory primate unit. In *Laboratory Animal Symposia*, Vol. lx, p. 113 (see Bibliography).

Cuthbertson, W. F. J. (1957). Nutritional requirements of rats and mice. In *Laboratory Animals Centre Collected Papers*, Vol. 5, p. 27 (see Bibliography): and *Proc. Nutrition Soc.* 16, 70.

Davey, D. G. (1959). Establishing and maintaining a colony of specific pathogen-free mice, rats and guinea-pigs. In *Laboratory Animals Centre Collected Papers*, Vol. 8. p. 17 (see Bibliography).

DeBock, C. A. and Peters, A. (1963). Effect of thalidomide on the development of the chick embryo. *Nature, Lond.* 199, 1204.

Department of Employment and Productivity & Ministry of Health and Social Services, Northern Ireland (1968). *Code of Practice for the Protection of Persons*

Exposed to Ionizing Radiations in Research and Teaching (2nd ed.). London: HMSO.

Dinsley, M. (1963). Inbreeding and selection. In *Animals for Research—Principles of Breeding and Management*, p. 235 (see Bibliography).

Drepper, K. (1967). Production of diets for SPF and germ-free animals with special regard to injury of the protein value following sterilization. In *Husbandry of Laboratory Animals*, p. 207 (see Bibliography).

Drepper, K. and Weik, H. (1970), Fütterung von Laboratoriumstieren in der Zucht und im Experiment. II Empfohlene Inhaltsstoffe in Standard- und Sonderdiäten für Mäuse, Ratten, Hamster, Meerschweinchen, Kaninchen, Hunde, Katzen und Affen. *Z. Versuchstierkunde* 12, 379.

Dubos, R. (1968). The gastrointestinal microbiota of the so-called normal mouse. *Carworth Europe Collected Papers* 2, 11.

Falconer, D. S. (1967). Genetic aspects of breeding methods. In *UFAW Handbook on the Care and Management of Laboratory Animals*, 3rd ed., p. 72 (see Bibliography).

Festing, M. F. W. and Grist, S. (1970). A simple technique for skin grafting rats. *Lab. Anim.* 4, 255.

Foster, H. L. (1959). Housing of disease-free vertebrates. *Ann. N.Y. Acad. Sci.* 78, 80.

Gledhill, A. W. (1962) Viral diseases in laboratory animals. In *The Problems of Laboratory Animal Disease*, p. 99 (see Bibliography).

Gärtner, K. (1969). Beziehung zwischen endokrinem System und soziologischen Situationen bei Massentierhaltung. *Zentbl. Vet Med.* 17, 81.

Grüneberg, H. (1955). Genetical aspects of breeding laboratory animals. In *Laboratory Animals Centre Collected Papers*, Vol. 3, p. 9 (see Bibliography).

Grüneberg, H. (1956). An annotated catalogue of the mutant genes of the house mouse. In *Medical Research Council Memorandum*, Vol. 33. London: HMSO.

Harris, J. M. (1965). Differences in responses between rat strains and colonies. *Fd Cosmet. Toxicol.* 3, 199.

Hill, A. V. (1929). Enemies of knowledge. Third Stephen Paget Memorial Lecture. *Fight Dis.* 17, 1.

Hill, B. F., Ed. (1963). Symposium on research animal housing. *Lab. Anim. Care* 13, No. 3, Pt 2, 219.

Hughes, D. (1968). Animal houses for radioactive work. In *Laboratory Animal Symposia*, Vol. 1, p. 119 (see Bibliography).

ILAR (1970). A nomenclatural system for outbred animals. Institute of Laboratory Animal Resources, Committee on Nomenclature. *Lab. Anim. Care* 20, 903.

IUCN (1966). *Red Data Book.* Morges, Switzerland: International Union for Conservation of Nature and Natural Resources: Survival Service Commission.

Jelínek, V. (1967). Vitamins in nutrition of laboratory animals. In *Husbandry of Laboratory Animals*, p. 97 (see Bibliography).

Joubert, C. J. (1967). Total nutritive requirements for small laboratory rodents (including rodents indigenous to South Africa). In *Husbandry of Laboratory Animals*, p. 133 (see Bibliography).

Kállai, L. (1971). Personal communication.

Kraft, L. M. (1967). Epidemiological aspects of disease in laboratory animals. In *Husbandry of Laboratory Animals*, p. 289 (see Bibliography).

Laboratory Animals Centre (1961). Hazards of the animal house. In *Laboratory Animals Centre Collected Papers*, Vol. 10 (see Bibliography).

Lane-Petter, M. E. (1970). Operating a mouse colony kept in filter racks. *J. Inst. Anim. Tech.* **21**, 158.

Lane-Petter, W. (1953). The accreditation scheme for laboratory animals. *Mon. Bull. Minist. Hlth* **12**, 165.

Lane-Petter, W. (1969). The grading of commercially-bred laboratory animals. *Vet. Rec.* **85**, 317.

Lane-Petter, W. (1970, a). A ventilation barrier to the spread of infection in laboratory animal colonies. *Lab. Anim.* **4**, 125.

Lane-Petter, W. (1970, b). Do laboratory animals like eating? *Proc. Nutn. Soc.* **29**, 335.

Lane-Petter, W. (1970, c). The filter rack: a device to protect laboratory animals against airborne infection. *Expn Anim.* **3**, 207.

Lane-Petter, W. (1970, d). The formulation of laboratory animal diets—some practical considerations. In *Nutrition and Disease in Experimental Animals*, p. 64 (see Bibliography).

Lane-Petter, W. (1970, e). Normal, hardy animals. *J. S. Afr. vet. med. Ass.* **41**, 149.

Lane-Petter, W. (1971). Dry and wet hysterectomy. *Z. Versuchstierkunde* **13**, 126.

Lane-Petter, W. and Bloom, J. L. (1957). Control of genetic variation. In *Laboratory Animals Centre Collected Papers*, Vol. 6, p. 51 (see Bibliography).

Lane-Petter, W., Brown, A. M., Cook, M. J., Porter, G. and Tuffery, A. A. (1959). Measuring productivity in breeding of small animals. *Nature, Lond.* **183**, 339.

Lane-Petter, W., Lane–Petter, M. E. and Bowtell, C. W. (1968). Intensive breeding of rats. I. Crossfostering. *Lab. Anim.* **2**, 35.

Lapage, G. (1960). *Achievement*. Cambridge: Heffer.

LASA (1969). *Dietary Standards for Laboratory Rats and Mice*. London: Laboratory Animals Ltd.

Ley, F. J., Bleby, J., Coates, M. E. and Paterson, J. S. (1969), Sterilization of laboratory animal diets using gamma radiation. *Lab. Anim.* **3**, 221.

Lyon, M. F. (1963). Genetics of the mouse. In *Animals for Research*, p. 199 (see Bibliography).

Medawar, P. B. (1957). Foreword to *Laboratory Animals Centre Collected Papers*, Vol. 6, p. 5 (see Bibliography).

Moutier, R. (1968). Détermination de quelques charactéristiques génétiques chez des souris de souche XLII/GIF. *Expn anim.* **1**, 261.

Moutier, R. (1971). Inbred strains of laboratory animals. In *Defining the Laboratory Animal*, p. 169 (see Bibliography).

Munro, J. W. (1966). *Pests of Stored Products*. London: Hutchinson.

NRC (1962). *Nutritional Requirements of Laboratory Animals*. Washington: Publications of the National Research Council, Vol. 990.

Nuffield Foundation (Division for Architectural Studies) (1961). *The Design of Research Laboratories*. London, New York, Toronto: Oxford University Press.

Ottewill, D. (1968). Laboratory animal houses. *Architect's J.* **147**, 1247; and Information Sheets, 1597–1602.

Parliament (1960). *Radioactive Substances Act*. London: HMSO.

Parrott, R. F. and Festing, M. F. W. (1971). *Standardised Laboratory Animals*. Laboratory Animals Centre Manual, Series No. 2.

Paterson, J. S. (1953). The provision of animals for research. In *Laboratory Animals Centre Collected Papers*, Vol. 1, p. 55 (see Bibliography).

Payne, P. R. (1967) The relationship between protein and calorie requirements of laboratory animals. In *Husbandry of Laboratory Animals*, p. 77 (see Bibliography).

Porter, G. (1968). Requirements of the animal. In *Laboratory Animal Symposia*, Vol. 1, p. 15 (see Bibliography).

Porter, G., and Bleby, J. (1966). Ethylene oxide sterilisation. Observations on the use of food, cages and nesting material so sterilised in the breeding of mice. *J. Inst. Anim. Tech.* 17, 160.

Porter, G. and Festing, M. (1970). A comparison between irradiated and auto-claved diets for breeding mice, with observations on palatability. *Lab. Anim.* 4, 203.

Porter, G., Lane-Petter, W. and Horne, N. (1963, a). Assessment of diets for mice. 1. Comparative feeding trials. 2. Diets in relation to reproduction. *Z. Versuchstierkunde* 2, 75 and 171.

Porter, G., Lane-Petter, W. and Horne, N. (1963, b). Effects of strong light on breeding mice. *J. Anim. Tech. Ass.* 16, 1.

Porter, G., Scott, P. P. and Walker, A. I. T. (1970). Caging standards for rats and mice: recommendations by the Laboratory Animal Science Association working party on caging and penning. *Lab. Anim.* 4, 61.

Quarterman, J. (1967). The importance of trace elements in the nutrition of laboratory animals. In *Husbandry of Laboratory Animals*, p. 115 (see Bibliography).

Quinn, E. H. and Pearson, A. E. G. (1968). Preliminary results from a cat breeding colony maintained under semi-external conditions. *J. Inst. Anim. Tech.* 19, 85.

Report (1962). *Safe Handling of Radioisotopes*. Vienna: International Atomic Energy Agency.

Report (1964, a). *Report of the Commission on Drug Safety*, p. 29. Washington: Federation of American Societies for Experimental Biology.

Report (1964, b). Terms and definitions. *ICLA Bulletin* 14, Annex 1.

Report (1965). *Report of the Departmental Committee on Experiments on Animals*. London: HMSO.

Report (1966). *Radiological Protection in Universities*. London: Vice-Chancellors' Committee, Association of Universities of the British Commonwealth.

Report (1969). *Lokaler för försöksdjur*. Stockholm: The National Swedish Board of Public Buildings.

Report (1970, a). Laboratory facilities and resources supporting biomedical research. *Lab. Anim. Care* 20, 795.

Report, (1970, b). Summary of questionnaire to determine number of animals used for research in 1969. *ILAR News* 14, No. 1 (i).

Reyniers, J. A., Sacksteder, M. R. and Ashburn, L. L. (1964). Multiple tumors in female germfree albino inbred mice exposed to bedding treated with ethylene oxide. *J. natn. Cancer Inst.* 32, 1045.

Ritchie, D. H. and Humphrey, J. K. (1970). Some observations on the mating of rats. *J. Inst. Anim. Tech.* 21, 100.

Sabourdy, M. (1970). Ed. *Les Mutants pathologiques chez l'animal*. Paris: Editions CNRS.

Sacquet, E. (1965). Pathogen-free animals. *Fd Cosmet. Toxicol.* **3**, 47.

Schneider, H. A. (1970). What can laboratory animal science gain from ecology? *ILAR News* **13**, No. 4, 2.

Scott, P. P. (1967). The cat. In *UFAW Handbook on the Care and Management of Laboratory Animals*, 3rd ed., p. 505 (see Bibliography).

Sellers, R. F. (1968). Infected animals. In *Laboratory Animal Symposia*, Vol. 1, p. 127 (see Bibliography).

Short, D. J. and Woodnott, D. P. (1969). In *The IAT Manual of Laboratory Animal Practice and Techniques*, Chs 10 and 22 (see Bibliography).

Staats, J. (1968). Standardized nomenclature for inbred strains of mice: fourth listing. *Cancer Res.* **28**, 391.

Statham, S. H. (1968). Materials for use in laboratory animal house construction. In *Laboratory Animal Symposia*, Vol. 1, p. 41 (see Bibliography).

Stewart, J. (1969). Uniformity and genetic variation in individuals and populations. *Carworth Europe Collected Papers* **3**. 51.

Thompson, G. P. (1960). The two aspects of science. *Nature, Lond.* **187**, 837.

Townsend, G. H. (1969). The grading of commercially-bred laboratory animals. *Veterinary Record* **85**, 225 and 420.

Tregier, A. and Homburger, F. (1961). Bacterial flora of the mouse uterus. *Proc. Soc. exp. Biol. Med.* **108**, 152.

Tuffery. A. A. (1959). The health of laboratory mice: a comparison of general health in two breeding units where different systems are employed. *J. Hyg., Camb.* **57**, 386.

Tuffery, A. A. (1962). Husbandry and health. In *Problems of Laboratory Animal Disease*, p. 239 (see Bibliography).

Van der Waaij, D. (1968). The persistent absence of Enterobacteriaceae from the intestinal flora of mice following antibiotic treatment. *J. infect. Dis.* **118**, 32.

Van der Waaij, D. and Sturm, C. A. (1968). Antibiotic decontamination of the digestive tract of mice. Technical procedures. *Lab. Anim. Care* **18**, 1

Walker, A. I. T. and Stevenson, D. E. (1967). The cost of building and running laboratory animal units. *Lab. Anim.* **1**, 105.

Weik, H. and Drepper, K. (1970). Fütterung von Laboratoriumstieren in der Zucht und im Experiment. II Empfohlene Inhaltsstoffe in Standard- und Sonderdiäten für Mäuse. Ratten. Hamster, Meerschweinchen, Kaninchen, Hunde, Katzen und Affen. *Z. Versuchstierkunde* **12**, 379.

Bibliography

This bibligraphy lists most of the books, published since 1950, to which those concerned with the breeding and care of laboratory animals will have frequent recourse. Although the majority are in the English language, some written in other languages are included. All are listed in alphabetical order of the first significant word of the title.

At the end of the bibliography is a list of periodical publications devoted to laboratory animal science and technology. Some of these are not widely known, but may contain papers of considerable importance. Some of them are the official journals of professional bodies listed in Chapter 11.

L'Animal de Laboratoire. M. Sabourdy (1967) (French). Paris: Presse Universitaires de France.
Good reference book for detailed treatment of genetics and breeder selection, and environmental and microbial effects on experimentation.

Animal Experiments for Medical Research. Eds. K. Ando and Y. Tajima (1956). *Medical Research and Animal Experiments.* K. Ando and T. Nomura (1960). Tokyo: Asakura Publishing Co.
These books are written in Japanese. The first is a general view of recent developments in laboratory animal science: in addition to the editors there are twenty-two co-authors. The second book is more philosophical in treatment.

Animals for Research—Principles of Breeding and Management. Ed. W. Lane-Petter (1963) (also in Czech). London, New York: Academic Press.
This book considers the commonest laboratory mammals, rats and mice, in greatest detail, but also includes chapters on guinea-pigs, hamsters, rabbits, cats, dogs, primates and chickens. Principles of management are discussed in some depth. The book is of particular interest to the research worker who needs to be led into a more profound understanding of his animals than the occasional user.

Les Animaux de Laboratoire. J. Dumas (1953) (French). Paris: Institut Pasteur.
This book deals mainly with disease and infections in common species of laboratory animals. It is far from comprehensive, but it contains information that is still of interest and is useful for reference.

Basic Care of Experimental Animals (revised 1968) (also in Greek and Spanish). New York: Animal Welfare Institute.
An elementary book, written in simple language for the student or the apprentice animal technician at the beginning of his career. The approach is very practical.

The Care and Breeding of Laboratory Animals. Ed. E. J. Farris (1950). New York: Wiley; London: Chapman and Hall.

269

One of the earliest books on laboratory animals. It is now largely out of date and has not been revised, but it is useful for reference.

Carworth Europe Collected Papers. Vols. 1–4 (1966–9). Huntingdon: Carworth Europe.
Vol. 1. *The Future of the Defined Laboratory Animal* (1967).
Vol. 2. *The Interaction of the Laboratory Animal with its Associated Organisms* (1968).
Vol. 3. *Uniformity* (1969).
Vol. 4. *The Effects of Environment* (1971).

Comfortable Quarters for Laboratory Animals (revised 1966). New York: Animal Welfare Institute.
Contains plans, photographs and some useful notes about many animal facilities, mostly in the USA, that the compilers have regarded as good examples of their kind.

Defining the Laboratory Animal in the Search for Health. (1971).
Proceedings of the fourth international symposium, organized jointly by the International Committee on Laboratory Animals and the Institute of Laboratory Animal Resources.

Experimental Surgical Preparations of Endocrine Glands. N. Simionescu and M. Scherzer (1969) (Romanian). Bucharest: Academic Publishing House.
A monograph on the methodology and technique of experimental surgery of the endocrine glands. It is classified by organs and generously illustrated.

Federation Proceedings. Published bimonthly by the Federation of American Societies for Experimental Biology.
(1960) *Laboratory Animals: an Annotated Bibliography.* Vol. 19. No. 4.
(1963) *Laboratory Animals: an Annotated Bibliography*, Supplement A. Vol. 22, No. 2 (Supplement No. 13).
(1969) *VA Laboratory Guide: Comparative Anaesthesia in Laboratory Animals.* Vol. 28, No. 4.
Three very useful reference sources.

Forsøksdyrbiologi. S. Erichsen (1969) (Norwegian). Oslo: Universitetsforlaget.
A general review of experimental animal research.

The Germ-Free Animal in Research. Ed. M. E. Coates (1968). London, New York: Academic Press.
A comprehensive review of germ-free techniques and applications, including contributions by most of the leading scientists in this field.

Husbandry of Laboratory Animals. Ed. M. L. Conalty (1967). Proceedings of the third international symposium organized by the International Committee on Laboratory Animals. London, New York: Academic Press.
This book covers a wide field. It contains sections on handling, production and organization; on nutrition; on ecology and disease; on zoonoses and other diseases; on physiology and psychophysiology; and on pharmacology and psycho-ecology. It must be regarded as a useful reference book.

The IAT Manual of Laboratory Animal Practice and Techniques. Ed. D. J. Short and
   D. P. Woodnott (2nd ed. 1969). London: Crosby Lockwood.
The subject matter of this well-established manual is based on the syllabuses for
the examinations of the Institute of Animal Technicians. The treatment is plain
and didactic. There is a lot of useful detailed information that cannot so easily be
found elsewhere.

International Index of Laboratory Animals (2nd Ed. 1971).
Compiled and distributed by the Laboratory Animals Centre, Carshalton, with
the assistance of the International Committee on Laboratory Animals.

International Symposium on Laboratory Animals. Eds. R. H. Regamey, W. Hennessen,
   D. Ikić and J. Ungar (1967). Basel, New York: Karger.
Proceedings of the sixteenth Symposium organized by the Permanent Section of
Microbiological Standardization. No. 5 in the series on Immunobiological
Standardization. Sessions on care and stock purity of laboratory animals; germ-
free, gnotobiotic and SPF animals; chronic infection and susceptibilities to
diseases; heritable pathological conditions, duplicate testing and reproductivity;
strains and species variations in pharmacological and biological responses.

Laboratory Animals Centre (Bureau) Collected Papers. Vols. 1–12 (1953–63).
   Carshalton: Laboratory Animal Centre.
Vol.   1. *On Supply and Husbandry of Laboratory Animals* (1953).
Vol.   2. *The Design of Animal Houses* (1954).
Vol.   3. *The Breeding of Laboratory Animals* (1955).
Vol.   4. *Infections in Laboratory Animals* (1956).
Vol.   5. *The Nutrition of Laboratory Animals* (1957).
Vol.   6. *Humane Technique in the Laboratory* (1957).
Vol.   7. *The Organization and Administration of an Animal Division* (1958).
Vol.   8. *Quality in Laboratory Animals* (1959).
Vol.   9. *The Provision of Animals for Cancer Research* (1960).
Vol. 10. *Hazards of the Animal House* (1961).
Vol. 11. *The Environment of Laboratory Animals* (1962).
Vol. 12. *The Choice of the Experimental Animal* (1963).

Laboratory Animal Handbooks, London: Laboratory Animals.
No. 1. *The Design and Function of Laboratory Animal Houses.* Eds. R. Hare and
   P. N. O'Donoghue (1968).
No. 2. *Dietary Standards for Laboratory Rats and Mice.* Eds. M. E. Coates,
   P. N. O'Donoghue, P. R. Payne and R. J. Ward (1969).
No. 3. *Transplanted Tumours of Rats and Mice: an Index of Tumours and Host
   Strains.* D. C. Roberts (1969).
No. 4. *Hazards of Handling Simians.* Eds. F. T. Perkins and P. N. O'Donoghue
   (1969).

Laboratory Animal Technology. K. L. Kovalevski (1958) (Russian). Moscow:
   Medgiz.

Laboratory Animals: Their Care and Their Facilities. Ed. J. A. D. Cooper (1960).
   The Journal of Medical Education, Vol. 35, No. 1.
Publication of a symposium on laboratory animals covering the aspects of design,
organization, disease, and quality and training of staff.

Living Animal Material for Biological Research. Ed. W. Lane-Petter (1958). Available from ICLA (for address of secretary see p. 252).

Manual for Laboratory Animal Technicians. Ed. G. R. Collins (1967). Joliet, Illinois: American Association for Laboratory Animal Science.
Outline of basic concepts, definitions, anatomy and physiology, genetics, reproduction and nutrition. Detailed treatment of the husbandry of most laboratory animal species. A useful training manual for animal technicians.

Methods of Animal Experimentation. Ed. W. I. Gay (three volumes, 1965, 1965, 1968). New York, London: Academic Press.
These books cover most of the more usual and many unusual experimental manipulations of laboratory animals. They are packed with practical detail.

Notes on the Law Relating to Experiments on Animals in Great Britain (The Act of 1876) (3rd ed. 1967). Issued by the Research Defence Society, London.
Essential reference for any licensee under the Act of 1876.

Nutrition and Disease in Experimental Animals. Ed. W. D. Tavernor (1970). London: Baillière, Tindall and Cassell.
Proceedings of a symposium organized by the British Small Animal Veterinary Association, the British Laboratory Animal Veterinary Association and the Laboratory Animal Science Association. The papers cover a wide field of the nutrition of laboratory animals, together with certain diseases of epidemic or zoonotic importance.

Pathologie der Laboratoriumstiere. P. Cohrs, R. Jaffé and H. Meessen (two volumes, 1958). Berlin: Springer.
A comprehensive treatise on laboratory animal pathology.

Pathology of Laboratory Rats and Mice. Eds. E. Cotchin and F. J. C. Roe (1967). Oxford, Edinburgh: Blackwell.
A fairly comprehensive treatment of the subject, containing much material not to be found elsewhere. Few bacterial diseases are described, but there are chapters on virus, helminth and fungal diseases.

Principles of Human Experimental Technique. W. M. S. Russell and R. L. Burch (1959). London: Methuen.
This book examines the principles of animal experimentation, especially from the humane point of view. It introduces some original concepts, which have important practical applications.

The Problems of Laboratory Animal Disease. Ed. R. J. C. Harris (1962). Proceedings of the second international symposium organized by the International Committee on Laboratory Animals, with the assistance of the National Institutes of Health (USA). London, New York: Academic Press.
This book contains papers on a wide variety of subjects. Some of the reported discussions are especially interesting.

Provision of Laboratory Animals for Research—A Practical Guide. W. Lane-Petter (1961) (also in Russian). Amsterdam: Elsevier.
A practical outline of the provision and maintenance of laboratory animals. Parts of this book have been freely drawn on for the present book.

The Rat in Laboratory Investigation (2nd ed. 1949). Eds. E. J. Farris and J. Q. Griffith. New York: Hafner.
A standard and essential reference book for all users of the rat as an experimental animal.

Reproduction and Breeding Techniques for Laboratory Animals. Ed. E. S. E. Hafez (1970). Philadelphia: Lea and Febiger.
A comparative survey of reproductive anatomy and physiology in common species of laboratory animals, with chapters on the breeding of individual species, and some useful appendices and general references.

Small Animal Anaesthesia. Ed. O. Graham-Jones (1964). Oxford, London: Pergamon Press.
Proceedings of a symposium organized by the British Small Animal Veterinary Association and the Universities Federation for Animal Welfare. Contains papers about anaesthesia in primates, ungulates, carnivores, rodents, lagomorphs, reptiles, amphibia and aquatic animals.

Symposium International sur l'Avenir des Animaux de Laboratoires (1967) (French), Lyon: Institut Mérieux.
Papers presented at a symposium held in 1966. Sections on the standardization of laboratory rodents; laboratory animals in general; breeding and use in the laboratory of the miniature pig and the dog; and breeding and use in the laboratory of the monkey.

The UFAW Handbook on the Care and Management of Laboratory Animals (3rd ed. 1967, 4th ed. 1971). Edinburgh, London: Livingstone.
The first and still the standard work of reference in this field. Most species of importance in the laboratory are covered, the less usual ones getting rather fuller treatment in proportion to their importance, because the commoner laboratory species are also well written up in other books. A general section covers aspects of animal care, and special sections include chapters on individual species. This book has a universal appeal, but it is particularly useful to scientists at the beginning of their careers in research, and to senior technicians.

Periodical Publications—
Journals Relating to Laboratory Animal Science

In compiling the following list the authors have drawn freely from the *ICLA Bulletin* No. 28, March 1971, which contains a large selection of relevant journal and book titles, including many of marginal interest or devoted entirely to a particular species.

Animaux de Laboratoire. Monthly (1964). Service des Publications du Centre National de la Recherche Scientifique, 3ème Bureau, 15 Quai Anatole France, Paris 7e.
A monthly bibliographical review of relevant publications grouped by species.

Bio-Medical Purview. Quarterly (1961). National Society for Medical Research, Rochester, Minn.

Biomdel Abstracts. Quarterly (1970). Agroinform, Attila ut. 93, Budapest, 1. Abstract journal published in English, concerned with all aspects of laboratory animal management.

Biotechniek. Ten issues per year. Biotechniek, Radiobiologische Instituut TNO, Lange Kleiweg 151, Rijswijk (ZH), Holland.
Journal of the Dutch Association of Animal Technicians.

Bulletin. Monthly (1949). National Society for Medical Research, 1330 Massachusetts Avenue NW, Washington, DC 20005, USA.
Scientific papers and topics concerned with experimental animals and related legislation.

Bulletin of the Institute of Animal Technicians (formerly *Bulletin of the Animal Technicians Association*). Monthly (1964). The Institute of Animal Technicians, 16 Beaumont Street, Oxford.
Contains information concerned with Institute affairs and meetings, and short technical articles.

Carworth Quarterly Letter. Quarterly (1946). Carworth, New City, Rockland County, NY 10956, USA.
Articles of general interest, relating particularly to small rodents.

Charles River Digest. Quarterly (1962). Charles River Breeding Laboratories, Wilmington, Mass. 01887, USA.
Articles of general interest relating particularly to small rodents.

Conquest. Annual (1950). Research Defence Society, 11 Chandos Street, London, W1M 9DE.
Distributed to members of the Society. Concerned with the ethical and legal aspects of animal experimentation.

Experimental Animals (formerly *Bulletin of Experimental Animals*). Quarterly (1952). Japan Experimental Animal Research Association, Institute of Medical Science, University of Tokyo, PO Takanama, Tokyo 108.
In Japanese with English contents and summaries.

Expérimentation Animale. Quarterly (1968). Vigot Frères, 23 Rue de l'École de Médecine, Paris 6e, France.
Publication in French and English of papers relating to all aspects of experimentation in laboratory animals.

ICLA Bulletin. Six-monthly (1957). International Committee on Laboratory Animals, National Institute of Public Health, Postuttak Oslo 1, Norway.
Articles of general interest, notices of symposia and conferences and details of ICLA activities.

ILAR News. Quarterly (1957). Institute of Laboratory Animal Resources, 2101 Constitution Avenue, Washington, DC 20418.
Contains abstracts of selected publications and book reviews, references on special topics, results of surveys and special investigations and details of meetings relevant to laboratory animal science.

Information Report of the Animal Welfare Institute. Quarterly (1951). PO Box 3492, Grand Central Station, New York, NY 10017.
Intended mainly for the lay public and concerned with the ethical considerations of animal experimentation—such as humane methods of killing. The Animal Welfare Institute also produce useful booklets on basic animal husbandry of value to inexperienced animal technicians.

Journal of the Institute of Animal Technicians (formerly *Journal of the Animal Technicians Association*). Quarterly (1950). 16 Beaumont St., Oxford.
Articles of scientific and technical interest.

Laboratory Animal Science (formerly *Laboratory Animal Care*). Six issues per year (1950). Williams and Wilkins Co., 428E Preston Street, Baltimore 2, Md, USA.
The first journal devoted exclusively to all aspects of the science and techniques associated with laboratory animals.

Laboratory Animals. Half-yearly (1967). Laboratory Animals Ltd, 7 Warwick Court, London, WC1.
The journal of the Laboratory Animals Science Association, containing papers, reviews and short communications on all aspects of laboratory animal science, technique and education.

LAC News Letter. Six-monthly. Laboratory Animals Centre, Woodmansterne Road, Carshalton, Surrey, England.
Information sheet for which no charge is made.

LAIS Bulletin. Six-monthly. The Indian Council of Medical Research, Cancer Research Institute, Tata Memorial Centre, Bombay-12, India.
Information sheet with articles of general interest.

Nytt fra Dyreavdelingen. Irregular. Animal Department, National Institute of Public Health, Postuttak Oslo 1, Norway.
An information sheet published irregularly in Norwegian, for which no charge is made.

Small Animal Literature Survey (included in the *Journal of Small Animal Practice*). Monthly (1961). Blackwell Scientific Publications, 5 Alfred Street, Oxford.
A monthly list of references relevant to experimentation in animals, compiled by the Laboratory Animals Centre, Carshalton.

Zeitschrift für Versuchstierkunde. Six issues per year (1961). Gustav Fischer Verlag, Jena, Germany.
Publication in German and English of papers related to laboratory animal science.

Zwierzęta Laboratoryjne. Annual or six-monthly (1963). Paustwowy Zaklad Wydawnictw Lekarskich, Warsawa, u. Dluga 38/40, Poland.
Publication in Polish of papers relating to laboratory animal science, with summaries in English.

Author Index

Numbers shown in italics are those pages at the end of the book on which references are listed.

278   AUTHOR INDEX

Subject Index